Physical Properties
of Foods

Physical Properties of Foods

Serpil Sahin and Servet Gülüm Sumnu

Middle East Technical University
Ankara, Turkey

Serpil Sahin
Department of Food Engineering
Middle East Technical University
Ankara, 06531
Turkey
serp@metu.edu.tr

Servet Gülüm Sumnu
Department of Food Engineering
Middle East Technical University
Ankara, 06531
Turkey
gulum@metu.edu.tr

Library of Congress Control Number: 2005937128

ISBN-10: 0-387-30780-X   e-ISBN 0-387-30808-3   Printed on acid-free paper.
ISBN-13: 978-0387-30780-0

Printed in the United States of America.   (TB/MVY)

9 8 7 6 5 4 3 2 1

springer.com

TO OUR PARENTS

SEMİHA-ŞEVKET SAHIN

&

EMİNE-ERDOĞAN SUMNU

Who have given us our roots

Preface

We have been teaching an undergraduate course of Physical Properties of Foods in Department of Food Engineering in Middle East Technical University for four years. We have had difficulty in finding a suitable textbook for undergraduate students since standard physical properties of foods books do not cover all the physical properties such as size, shape, volume and related physical attributes, rheological properties, thermal properties, electromagnetic properties, water activity and sorption properties, and surface properties together. In addition, engineering concepts and physical chemistry knowledge are not combined in these books.

We tried to write a book to provide a fundamental understanding of physical properties of foods. In this book, the knowledge of physical properties is combined with food science, physical chemistry, physics and engineering knowledge. Physical properties data are required during harvesting, processing, storage and even shipping to the consumer. The material in the book will be helpful for the students to understand the relationship between physical and functional properties of raw, semi-finished and processed food to obtain products with desired shelf-life and quality.

This book discusses basic definitions and principles of physical properties, the importance of physical properties in food industry and measurement methods. Moreover, recent studies in the area of physical properties are summarized. In addition, each chapter is provided with examples and problems. These problems will be helpful for students for their self-study and to gain information how to analyze experimental data to generate practical information.

This book is written to be a textbook for undergraduate students which will fill the gap in physical properties area. In addition, the material in the book may be of interest to people who are working in the field of Food Science, Food Technology, Biological Systems Engineering, Food Process Engineering and Agricultural Engineering. It will also be helpful for graduate students who deal with physical properties in their research.

The physical properties of food materials are discussed in 6 main categories such as size, shape, volume and related physical attributes, rheological properties, thermal properties, electromagnetic properties, water activity and sorption properties and surface properties in this book. In the first chapter physical attributes of foods which are size, shape, volume, density and porosity are discussed. Methods to measure these properties are explained in details. In Chapter 2, after making an introduction on Newtonian and non-Newtonian fluid flow, viscosity measurement methods are discussed. Then, principle of viscoelastic fluids, methods to determine the viscoelastic behavior and models used in viscoelastic fluids are mentioned. Chapter 3 explains definition and measurement methods of thermal properties such as thermal conductivity, specific heat, thermal diffusivity and enthalpy. In the fourth chapter color and dielectric properties of foods are covered. In Chapter 5, equilibrium criteria and colligative properties are discussed. Then, information is given on measurement of water activity. Finally preparation of sorption isotherms and models are discussed. The last chapter is about surface properties and their measurement methods. Where appropriate, we have cited throughout the text the articles that are available for more information.

We are deeply grateful to Prof. Dr. Haluk Hamamci for encouraging us during writing this book and his belief in us. We would also like to thank our colleagues Prof. Dr. Ali Esin, Prof. Dr. Haluk Hamamci, Assoc. Prof. Dr. Nihal Aydogan, Assoc. Prof. Dr. Pinar Calik, Assoc. Prof. Dr. Naime Asli Sezgi, Assoc. Prof. Dr. Esra Yener, Assist. Prof. Dr. Yusuf Uludag, who read the chapters and gave useful suggestions. We would also like to thank our Ph.D student Halil Mecit Oztop and his brother Muin S. Oztop for drawing some of the figures. We would like to extend our thanks to our Ph.D students, Isil Barutcu, Suzan Tireki, Semin Ozge Keskin, Elif Turabi and Nadide Seyhun for reviewing our book. We are happy to acknowledge the teaching assistants Aysem Batur and Incinur Hasbay for their great effort in drawing some of the figures, finding the examples and problems given in each chapter.

Last but not the least; we would like to thank our families for their continuous support throughout our academic career. With love, this work is dedicated to our parents who have patience and belief in us. Thank you for teaching us how to struggle the difficulties in life.

Ankara-TURKEY   Serpil Sahin
October 29, 2005   Servet Gülüm Sumnu

Contents

Size, Shape, Volume, and Related Physical Attributes

SUMMARY

In this chapter, the physical attributes of foods, which consist of size, shape, volume, density, and porosity, are discussed. Methods to measure these properties are explained in detail.

Size and shape are important physical attributes of foods that are used in screening, grading, and quality control of foods. They are also important in fluid flow and heat and mass transfer calculations. Sieve analysis can be used to determine the average particle diameter and specific surface area of granular material. Volume, which affects consumer acceptance, can be calculated from the measured dimensions or by using various methods such as liquid, gas, or solid displacement methods and image processing. Volume measurement methods can also be used for measuring the density of solids. Volume/density can be expressed in different forms such as solid, apparent, and bulk volume/density depending on pores. Porosity is a physical property characterizing the texture and the quality of dry and intermediate moisture foods. Total porosity of particulate materials includes the voids within and among the particles. Porosity can be determined from the difference between bulk volume of a piece of porous material and its volume after destruction of all voids by compression, optical methods, density methods, or by using a pycnometer or porosimeter. Internal pores may be in three different forms: closed pores that are closed on all sides, blind pores in which one end is closed, and flow-through pores that are open at both ends so that flow can take place.

1.1 SIZE

Size is an important physical attribute of foods used in screening solids to separate foreign materials, grading of fruits and vegetables, and evaluating the quality of food materials. In fluid flow, and heat and mass transfer calculations, it is necessary to know the size of the sample. Size of the particulate foods is also critical. For example, particle size of powdered milk must be large enough to prevent agglomeration, but small enough to allow rapid dissolution during reconstitution. Particle size was found to be inversely proportional to dispersion of powder and water holding capacity of whey protein powders (Resch & Daubert, 2001). Decrease in particle size also increased the steady shear and

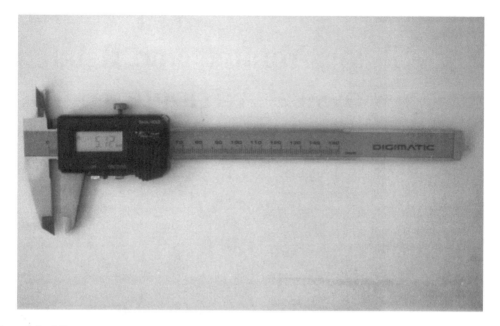

Figure 1.1 Micrometer.

complex viscosity of the reconstituted powder. The powder exhibited greater intrinsic viscosity as the particle size increased. The size of semolina particles was found to influence mainly sorption kinetics (Hebrard, Oulahna, Galet, Cuq, Abecassis, & Fages, 2003). The importance of particle size measurement has been widely recognized, especially in the beverage industry, as the distribution and concentration ratio of particulates present in beverages greatly affect their flavor.

It is easy to specify size for regular particles, but for irregular particles the term size must be arbitrarily specified.

Particle sizes are expressed in different units depending on the size range involved. Coarse particles are measured in millimeters, fine particles in terms of screen size, and very fine particles in micrometers or nanometers. Ultrafine particles are sometimes described in terms of their surface area per unit mass, usually in square meters per gram (McCabe, Smith & Harriot, 1993).

Size can be determined using the projected area method. In this method, three characteristic dimensions are defined:

1. Major diameter, which is the longest dimension of the maximum projected area;
2. Intermediate diameter, which is the minimum diameter of the maximum projected area or the maximum diameter of the minimum projected area; and
3. Minor diameter, which is the shortest dimension of the minimum projected area.

Length, width, and thickness terms are commonly used that correspond to major, intermediate, and minor diameters, respectively.

The dimensions can be measured using a micrometer or caliper (Fig. 1.1). The micrometer is a simple instrument used to measure distances between surfaces. Most micrometers have a frame, anvil, spindle, sleeve, thimble, and ratchet stop. They are used to measure the outside diameters, inside diameters, the distance between parallel surfaces, and the depth of holes.

Particle size of particulate foods can be determined by sieve analysis, passage through an electrically charged orifice, and settling rate methods. Particle size distribution analyzers, which determine both the size of particles and their state of distribution, are used for production control of powders.

1.2 SHAPE

Shape is also important in heat and mass transfer calculations, screening solids to separate foreign materials, grading of fruits and vegetables, and evaluating the quality of food materials. The shape of a food material is usually expressed in terms of its sphericity and aspect ratio.

Sphericity is an important parameter used in fluid flow and heat and mass transfer calculations. Sphericity or shape factor can be defined in different ways.

According to the most commonly used definition, sphericity is the ratio of volume of solid to the volume of a sphere that has a diameter equal to the major diameter of the object so that it can circumscribe the solid sample. For a spherical particle of diameter D_p, sphericity is equal to 1 (Mohsenin, 1970).

$$\text{Sphericity} = \left(\frac{\text{Volume of solid sample}}{\text{Volume of circumscribed sphere}} \right)^{1/3} \tag{1.1}$$

Assuming that the volume of the solid sample is equal to the volume of the triaxial ellipsoid which has diameters equivalent to those of the sample, then:

$$\Phi = \left(\frac{V_e}{V_c} \right)^{1/3} \tag{1.2}$$

where

Φ = sphericity,
V_e = volume of the triaxial ellipsoid with equivalent diameters (m^3),
V_c = volume of the circumscribed sphere (m^3).

In a triaxial ellipsoid, all three perpendicular sections are ellipses (Fig. 1.2). If the major, intermediate, and minor diameters are $2a$, $2b$, and $2c$, respectively, volume of the triaxial ellipsoid can be determined from the following equation:

$$V_e = \frac{4}{3} \pi abc \tag{1.3}$$

Then, sphericity is:

$$\Phi = \frac{(abc)^{1/3}}{a} \tag{1.4}$$

Sphericity is also defined as the ratio of surface area of a sphere having the same volume as the object to the actual surface area of the object (McCabe, Smith, & Harriot, 1993):

$$\Phi = \frac{\pi D_p^2}{S_p} = \frac{6V_p}{D_p S_p} \tag{1.5}$$

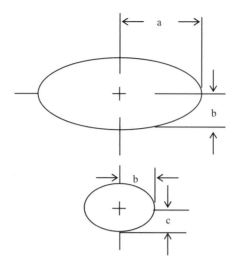

Figure 1.2   Triaxial ellipsoid

where

D_p = equivalent diameter or nominal diameter of the particle (m),
S_p = surface area of one particle (m^2),
V_p = volume of one particle (m^3).

The equivalent diameter is sometimes defined as the diameter of a sphere having the same volume as the particle. However, for fine granular materials, it is difficult to determine the exact volume and surface area of a particle. Therefore, equivalent diameter is usually taken to be the nominal size based on screen analysis or microscopic examination in granular materials. The surface area is found from adsorption measurements or from the pressure drop in a bed of particles.

In general, diameters may be specified for any equidimensional particle. Particles that are not equidimensional, that is, longer in one direction than in others, are often characterized by the second longest major dimension. For example, for needlelike particles, equivalent diameter refers to the thickness of the particles, not their length.

In a sample of uniform particles of diameter D_p, the number of particles in the sample is:

$$N = \frac{m}{\rho_p V_p} \tag{1.6}$$

where

N = the number of particles,
m = mass of the sample (kg),
ρ_p = density of the particles (kg/m^3),
V_p = volume of one particle (m^3).

Total surface area of the particles is obtained from Eqs. (1.5) and (1.6):

$$A = N S_p = \frac{6m}{\Phi \rho_p D_p} \tag{1.7}$$

Another definition of sphericity is the ratio of the diameter of the largest inscribed circle (d_i) to the diameter of the smallest circumscribed circle (d_c) (Mohsenin, 1970):

$$\Phi = \frac{d_i}{d_c} \tag{1.8}$$

Recently, Bayram (2005) proposed another equation to calculate sphericity as:

$$\Phi = \frac{\sum \left(D_i - \overline{D}\right)^2}{\left(\overline{D}N\right)^2} \tag{1.9}$$

where

D_i = any measured dimension (m),
$\overline{D}$ = average dimension or equivalent diameter (m),
N = number of measurements (the increase in N increases the accuracy).

According to this formula, equivalent diameter for irregular shape material is accepted as the average dimension. Differences between average diameter and measured dimensions are determined by the sum of square of differences. When this difference is divided by the square of product of the average diameter and number of measurements, it gives a fraction for the approach of the slope to an equivalent sphere, which is sphericity.

According to Eq. (1.9), if the sample sphericity value is close to zero it can be considered as spherical. Table 1.1 shows sphericity values of some granular materials determined by Eq. (1.9).

Example 1.1. Calculate the sphericity of a cylindrical object of diameter 1.0 cm and height 1.7 cm.

Solution:

The volume of the object can be calculated by,

$$V = \pi r^2 h = \pi(0.5)^2(1.7) = 1.335 \text{ cm}^3$$

Table 1.1   Sphericity Values for Granular Materials

Type of Product	Φ
Wheat	0.01038
Bean	0.00743
Intact red lentil	0.00641
Chickpea	0.00240
Coarse bulgur	0.01489

From Bayram (2005).

The radius of the sphere (r_s) having this volume can be calculated as:

$$\frac{4}{3}\pi r_s^3 = 1.335 \text{ cm}^3$$

$$\Rightarrow r_s = 0.683 \text{ cm}$$

The surface area of sphere of the same volume as the particle is:

$$S_s = 4\pi r_s^2 = 4\pi (0.683)^2 = 5.859 \text{ cm}^2$$

The surface area of the particle is:

$$S_p = 2\pi r(h + r) = 2\pi (0.5)(1.7 + 0.5) = 6.908 \text{ cm}^2$$

Then, sphericity is calculated as:

$$\Phi = \frac{S_s}{S_p} = \frac{5.859}{6.908} = 0.848$$

The aspect ratio (R_a) is another term used to express the shape of a material. It is calculated using the length (a) and the width (b) of the sample as (Maduako & Faborode, 1990):

$$R_a = \frac{b}{a} \tag{1.10}$$

Certain parameters are important for the design of conveyors for particulate foods, such as radius of curvature, roundness, and angle of repose. **Radius of curvature** is important to determine how easily the object will roll. The more sharply rounded the surface of contact, the greater will be the stresses developed. A simple device for measuring the radius of curvature is shown in Fig. 1.3. It consists of a metal base that has a dial indicator and holes into which pins are placed. Two pins are placed within these holes according to the size of the object. When the two pins make contact with the surface, the tip of the dial indicator is pushed upwards. Then, the dial indicator reads the sagittal height (S). The radius of curvature is calculated from the measured distances using this simple device and the following formula:

$$\text{Radius of curvature} = \frac{(D/2)^2 + S^2}{2S} \tag{1.11}$$

where

D = spacing between the pins (m),
S = sagittal height (m).

The minimum and the maximum radii of curvature for larger objects such as apples are calculated using the larger and smaller dial indicator readings, respectively.

For smaller objects of relatively uniform shape, the radius of curvature can be calculated using the major diameter and either the minor or intermediate diameter.

$$R_{\min} = \frac{H}{2} \tag{1.12}$$

$$R_{\max} = \frac{H^2 + \dfrac{L^2}{4}}{2H} \tag{1.13}$$

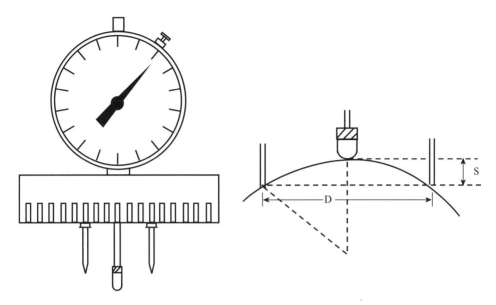

Figure 1.3   A device for measuring the radius of curvature.

where

R_{min} = Minimum radius of curvature (m),
R_{max} = Maximum radius of curvature (m),
H = intermediate diameter or the average of minor and major diameters (m),
L = major diameter (m).

Example 1.2. The major diameter (L) and the average of the minor and major diameters (H) of barley are measured as 8.76 mm and 2.83 mm, respectively. Calculate the minimum and maximum radii of curvature for the barley.

Solution:

The minimum and maximum radius of curvatures can be calculated using Eqs. (1.12) and (1.13), respectively:

$$R_{min} = \frac{H}{2} = \frac{2.83}{2} = 1.415 \text{ mm} \tag{1.12}$$

$$R_{max} = \frac{H^2 + \dfrac{L^2}{4}}{2H} = \frac{(2.83)^2 + \dfrac{(8.76)^2}{4}}{2(2.83)} = 4.804 \text{ mm} \tag{1.13}$$

Roundness is a measure of the sharpness of the corners of the solid. Several methods are available for estimating roundness. The most commonly used ones are given below (Mohsenin, 1970):

$$\text{Roundness} = \frac{A_p}{A_c} \tag{1.14}$$

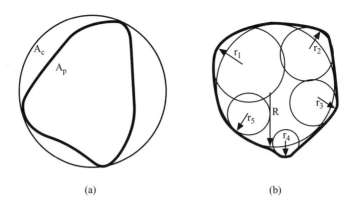

(a)   (b)

Figure 1.4   Roundness definitions.

where

A_p = largest projected area of object in natural rest position (m^2),
A_c = Area of the smallest circumscribing circle as defined in Fig. 1.4a (m^2).

Roundness can also be estimated from Eq. (1.15):

$$\text{Roundness} = \frac{\sum_{i=1}^{N} r}{NR} \qquad (1.15)$$

where

r = radius of curvature as defined in Fig. 1.4b (m),
R = radius of the maximum inscribed circle (m),
N = total number of corners summed in numerator.

Angle of repose is another important physical property used in particulate foods such as seeds, grains, and fruits. When granular solids are piled on a flat surface, the sides of the pile are at a definite reproducible angle with the horizontal. This angle is called the angle of repose of the material. The angle of repose is important for the design of processing, storage, and conveying systems of particulate material. When the grains are smooth and rounded, the angle of repose is low. For very fine and sticky materials the angle of repose is high. For determination of this property, a box with open sides at the top and bottom is placed on a surface. The angle of repose is determined by filling the box with sample and lifting up the box gradually, allowing the sample to accumulate and form a conical heap on the surface. Then, the angle of repose is calculated from the ratio of the height to the base radius of the heap formed.

1.3   PARTICLE SIZE DISTRIBUTION

The range of particle size in foods depends on the cell structure and the degree of processing. The hardness of grain is a significant factor in the particle size distribution of flour. The particle size distribution of flour is known to play an important role in its functional properties and the quality of end products. The relationship between the physicochemical properties of rice grains and particle size distributions of rice flours from different rice cultivars were examined (Chen, Lii, & Lu, 2004). It was found that physical characteristics of rice grain were the major factors but chemical compositions were also important in affecting the particle size distribution of rice flour.

To apply Eqs. (1.6) and (1.7) to a mixture of particles having various sizes and densities, the mixture is sorted into fractions, each of constant density and approximately constant size. Each fraction can be weighed or the individual particles in it can be counted. Then, Eqs. (1.6) and (1.7) can be applied to each fraction and the results can be added.

Particles can be separated into fractions by using one of the following methods:

1. Air elutriation method: In this method, the velocity of an air stream is adjusted so that particles measuring less than a given diameter are suspended. After the particles within the size range are collected, the air velocity is increased and the new fraction of particles is collected. The process continues until the particulate food is separated into different fractions.

2. Settling, sedimentation, and centrifugation method: In settling and sedimentation, the particles are separated from the fluid by gravitational forces acting on the particles. The particles can be solid particles or liquid drops. Settling and sedimentation are used to remove the particles from the fluid. It is also possible to separate the particles into fractions of different size or density. Particles that will not settle by gravitational force can be separated by centrifugal force. If the purpose is to separate the particles into fractions of different sizes, particles of uniform density but different sizes are suspended in a liquid and settle at different rates. Particles that settle in given time intervals are collected and weighed.

3. Screening: This is a unit operation in which various sizes of solid particles are separated into two or more fractions by passing over screen(s). A dispersing agent may be added to improve sieving characteristics. Screen is the surface containing a number of equally sized openings. The openings are square. Each screen is identified in meshes per inch. Mesh is defined as open spaces in a network. The smallest mesh means largest clear opening.

A set of standard screens is stacked one upon the other with the smallest opening at the bottom and the largest at the top placed on an automatic shaker for screen analysis (sieve analysis). In screen analysis, the sample is placed on the top screen and the stack is shaken mechanically for a definite time. The particles retained on each screen are removed and weighed. Then, the mass fractions of particles separated are calculated. Any particles that pass through the finest screen are collected in a pan at the bottom of the stack.

Among the standard screens, the Tyler Standard Screen Series is the most commonly used sieve series (Table 1.2). The area of openings in any screen in the series is exactly twice the openings in the next smaller screen. The ratio of actual mesh dimension of any screen to that of the next smaller screen is $\sqrt{2} = 1.41$. For closer sizing, intermediate screens are available that have mesh dimension $\sqrt[4]{2} = 1.189$ times that of the next smaller standard screen.

Since the particles on any one screen are passed by the screen immediately ahead of it, two numbers are required to specify the size range of an increment: one for the screen through which the fraction passes and the other on which it is retained. For example, 6/8 refers to the particles passing through the 6-mesh and remaining on an 8-mesh screen.

Particle size analysis can be done in two different ways: differential analysis and cumulative analysis.

In differential analysis, mass or number fraction in each size increment is plotted as a function of average particle size or particle size range. The results are often presented as a histogram as shown in Fig. 1.5 with a continuous curve to approximate the distribution. If the particle size ranges are all equal

Table 1.2 Tyler Standard Screen Scale

Mesh	Clear Opening (in.)	Clear Opening (mm)
$2\frac{1}{2}$[a]	0.312	7.925
3	0.263	6.68
$3\frac{1}{2}$[a]	0.221	5.613
4	0.185	4.699
5[a]	0.156	3.962
6	0.131	3.327
7[a]	0.110	2.794
8	0.093	2.362
9[a]	0.078	1.981
10	0.065	1.651
12[a]	0.055	1.397
14	0.046	1.168
16[a]	0.039	0.991
20	0.0328	0.833
24[a]	0.0276	0.701
28	0.0232	0.589
32[a]	0.0195	0.495
35	0.0164	0.417
42[a]	0.0138	0.351
48	0.0116	0.295
60[a]	0.0097	0.246
65	0.0082	0.208
80[a]	0.0069	0.175
100	0.0058	0.147
115[a]	0.0049	0.124
150	0.0041	0.104
170[a]	0.0035	0.088
200	0.0029	0.074

[a]These screens, for closer sizing, are inserted between the sizes usually considered as the standard series. With the inclusion of these screens the ratio of diameters of openings in two successive screens is $1:\sqrt[4]{2}$ instead of $1:\sqrt{2}$.

as in this figure, the data can be plotted directly. However, it gives a false impression if the covered range of particle sizes differs from increment to increment. Less material is retained in an increment when the particle size range is narrow than when it is wide. Therefore, average particle size or size range versus $X_i^w / D_{pi+1} - D_{pi}$ should be plotted, where X_i^w is the mass fraction and $\left(D_{pi+1} - D_{pi}\right)$ is the particle size range in increment i (McCabe et al., 1993).

Cumulative analysis is obtained by adding, consecutively, the individual increments, starting with that containing the smallest particles and plotting the cumulative sums against the maximum particle diameter in the increment. In a cumulative analysis, the data may appropriately be represented by a continuous curve. Table 1.3 shows a typical screen analysis. Cumulative plots are made using the second and fifth columns of Table 1.3 (Fig. 1.6).

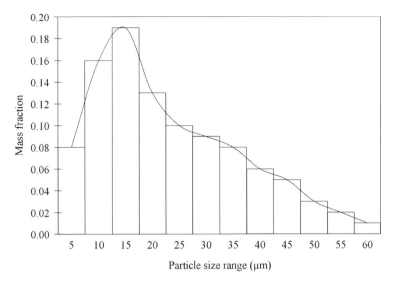

Figure 1.5 Particle size distributions using differential analysis.

Calculations of average particle size, specific surface area, or particle population of a mixture may be based on either a differential or a cumulative analysis. In cumulative analysis, the assumption of "all particles in a single fraction are equal in size" is not required. Therefore, methods based on the cumulative analysis are more precise than those based on differential analysis.

Table 1.3 A Typical Screen Analysis

Mesh	Screen Opening D_{pi} (mm)	Mass fraction Retained (X_i^w)	Average Diameter $\overline{D}_{pi}$ (mm)	Percentage Smaller Than D_{pi}
28	0.589	0	—	100
32	0.495	0.08	0.5420	92
35	0.417	0.06	0.4560	86
42	0.351	0.08	0.3840	78
48	0.295	0.14	0.3230	64
60	0.246	0.14	0.2705	50
65	0.208	0.14	0.2270	36
80	0.175	0.09	0.1915	27
100	0.147	0.09	0.1610	18
115	0.124	0.06	0.1355	12
150	0.104	0.04	0.1140	8
170	0.088	0.04	0.0960	4
200	0.074	0.02	0.0810	2
Pan	—	0.02	0.0370	0

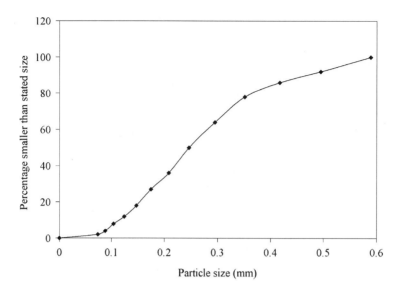

Figure 1.6   Particle size distributions using cumulative analysis.

If the particle density and sphericity are known, the surface area of the particles in each fraction may be calculated from Eq. (1.7) and the results for each fraction are added to give the specific surface area of mixture. The specific surface area is defined as the total surface area of a unit mass of particles. For constant density (ρ_p) and sphericity (Φ), specific surface area (A_w) of the mixture is:

$$A_w = \frac{6}{\Phi \rho_p} \sum_{i=1}^{n} \frac{X_i^w}{\overline{D}_{pi}} \tag{1.16}$$

where

i = subscript showing individual increments,
X_i^w = mass fraction in a given increment,
n = number of increments,
$\overline{D}_{pi}$ = average particle diameter taken as the arithmetic mean of the smallest and largest particle diameters in the increment (m) and expressed as

$$\overline{D}_{pi} = \frac{D_{pi} + D_{p(i-1)}}{2} \tag{1.17}$$

If the cumulative analysis is used, specific surface area of the mixture is found by integrating with respect to mass fraction between the limits of 0 to 1 (McCabe & Smith, 1976):

$$A_w = \frac{6}{\Phi \rho_p} \int_0^1 \frac{dX^w}{D_p} \tag{1.18}$$

Average particle diameter of a mixture can be calculated in different ways. The most commonly used one is the volume surface mean diameter (Sauter mean diameter). It is used if the mass fraction of particles in each fraction is known. For differential analysis:

$$\overline{D_s} = \frac{1}{\sum\limits_{i=1}^{n} \dfrac{X_i^w}{\overline{D_{pi}}}} \tag{1.19}$$

For cumulative analysis:

$$\overline{D_s} = \frac{1}{\int\limits_{0}^{1} \dfrac{dX^w}{D_p}} \tag{1.20}$$

Mass mean diameter can also be calculated if the mass fractions of particles in each fraction are known. For differential analysis:

$$\overline{D_w} = \sum_{i=1}^{n} \overline{D_{pi}} X_i^w \tag{1.21}$$

For cumulative analysis:

$$\overline{D_w} = \int_{0}^{1} D_p dX^w \tag{1.22}$$

If the number of particles in each fraction is known, arithmetic mean diameter is used. For differential analysis:

$$\overline{D_N} = \frac{\sum\limits_{i=1}^{n} \overline{D}_{pi} N_i}{N_T} \tag{1.23}$$

where

N_i = number of particles in each fraction,
N_T = total number of particles,
n = number of size groups.

For cumulative analysis:

$$\overline{D_N} = \frac{\int\limits_{0}^{1} D_p dN}{N_T} \tag{1.24}$$

The number of particles in the mixture can be calculated from either differential or cumulative analysis using Eqs. (1.25) and (1.26), respectively:

$$N_T = \frac{1}{\varphi \rho_p} \sum_{i=1}^{n} \frac{X_i^w}{D_{pi}^3} \tag{1.25}$$

$$N_T = \frac{1}{\varphi \rho_p} \int_{0}^{1} \frac{dX^w}{D_p^3} \tag{1.26}$$

where φ is the volume shape factor, which is defined by the ratio of volume of a particle (V_p) to its cubic diameter:

$$\varphi = \frac{V_p}{D_p^3} \tag{1.27}$$

Dividing the total volume of the sample by the number of particles in the mixture gives the average volume of a particle. The diameter of such a particle is the volume mean diameter, which is found from:

$$\overline{D_V} = \left[\frac{1}{\displaystyle\sum_{i=1}^{n} \frac{X_i^w}{(\overline{D}_{pi})^3}} \right]^{1/3} \tag{1.28}$$

For the cumulative analysis, volume mean diameter is determined by integrating with respect to mass fraction between the limits of 0 and 1:

$$\overline{D_v} = \left[\frac{1}{\displaystyle\int_0^1 \frac{dX^w}{D_p^3}} \right]^{1/3} \tag{1.29}$$

Example 1.3. Wheat flour is made by grinding the dry wheat grains. Particle size is an important characteristic in many of the wheat products. For example, in making wafers, if the flour is too fine, light and tender products are formed. On the other hand, incomplete sheets of unsatisfactory wafers are formed if the flour is too coarse. Therefore, it is important to test the grinding performance of flour by sieve analysis in wafer producing factories. Determine the volume surface mean diameter, mass mean diameter, and volume mean diameter of wheat flour by differential analysis using the data given in Table E.1.3.1.

Table E.1.3.1 Sieve Analysis of Wheat Flour

Mesh	Quantity of Wheat Flour Retained (g)
8/10	1.98
10/20	11.69
20/32	4.51
32/42	1.20
42/60	2.43
60/80	0.63
80/100	0.69
100/Pan	1.47

Table E.1.3.2 Differential Analysis
of Wheat Flour

Mesh	Mass Fraction Retained	$\overline{D}_{pi}$
8/10	0.080	2.007
10/20	0.475	1.242
20/32	0.183	0.664
32/42	0.049	0.423
42/60	0.099	0.299
60/80	0.026	0.211
80/100	0.028	0.161
100/Pan	0.060	0.074

Solution:

The total mass of wheat flour is determined by adding the quantity of wheat flour retained on each screen having different mesh:

$$m_{\text{total}} = 1.98 + 11.69 + 4.51 + 1.20 + 2.43 + 0.63 + 0.69 + 1.47 = 24.60 \text{ g}$$

Using Table 1.2 and Eq. (1.17), $\overline{D}_{pi}$ is calculated and differential analysis results are tabulated (Table E.1.3.2).

$$\overline{D}_{pi} = \frac{D_{pi} + D_{p(i-1)}}{2} \tag{1.17}$$

Then, volume surface mean diameter, mass mean diameter, and volume mean diameter are calculated using Eqs. (1.19), (1.21), and (1.28), respectively.

$$\overline{D}_s = \frac{1}{\sum\limits_{i=1}^{n} \dfrac{X_i^w}{\overline{D}_{pi}}} = 0.444 \text{ mm} \tag{1.19}$$

$$\overline{D}_w = \sum\limits_{i=1}^{n} \overline{D}_{pi} X_i^w = 0.938 \text{ mm} \tag{1.21}$$

$$\overline{D}_V = \left[\frac{1}{\sum\limits_{i=1}^{n} \dfrac{X_i^w}{\left(\overline{D}_{pi}\right)^3}} \right]^{1/3} = 0.182 \text{ mm} \tag{1.28}$$

1.4 VOLUME

Volume is defined as the amount of three-dimensional space occupied by an object, usually expressed in units that are the cubes of length units, such as cubic inches and cubic centimeters, or in units of liquid measure, such as gallons and liters. In the SI system, the unit of volume is m^3.

Volume is an important quality attribute in the food industry. It appeals to the eye, and is related to other quality parameters. For instance, it is inversely correlated with texture.

Volume of solids can be determined by using the following methods:

1. Volume can be calculated from the characteristic dimensions in the case of objects with regular shape.
2. Volumes of solids can be determined experimentally by liquid, gas, or solid displacement methods.
3. Volume can be measured by the image processing method. An image processing method has been recently developed to measure volume of ellipsoidal agricultural products such as eggs, lemons, limes, and peaches (Sabliov, Boldor, Keener, & Farkas, 2002).

Liquid, gas, and solid displacement methods are described in the following sections.

1.4.1   Liquid Displacement Method

If the solid sample does not absorb liquid very fast, the liquid displacement method can be used to measure its volume. In this method, volume of food materials can be measured by pycnometers (specific gravity bottles) or graduated cylinders. The pycnometer has a small hole in the lid that allows liquid to escape as the lid is fitted into the neck of the bottle (Fig. 1.7). The bottle is precisely weighed and filled with a liquid of known density. The lid is placed on the bottle so that the liquid is forced out of the capillary. Liquid that has been forced out of the capillary is wiped from the bottle and the bottle is weighed again. After the bottle is emptied and dried, solid particles are placed in the bottle and the bottle is weighed again. The bottle is completely filled with liquid so that liquid is forced from the hole when the lid is replaced. The bottle is reweighed and the volume of solid particles can be

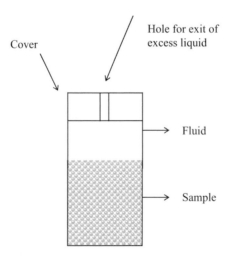

Figure 1.7   Pycnometer (specific gravity bottle).

determined from the following formula:

$$V_s = \frac{\text{Weight of the liquid displaced by solid}}{\text{Density of liquid}}$$

$$= \frac{(W_{pl} - W_p) - (W_{pls} - W_{ps})}{\rho_l} \tag{1.30}$$

where

V_s = volume of the solid (m^3),
W_{pl} = weight of the pycnometer filled with liquid (kg),
W_p = weight of the empty pycnometer (kg),
W_{pls} = weight of the pycnometer containing the solid sample and filled with liquid (kg),
W_{ps} = weight of the pycnometer containing solid sample with no liquid (kg),
ρ_l = density of the liquid (kg/m^3).

The volume of a sample can be measured by direct measurement of volume of the liquid displaced by using a graduated cylinder or burette. The difference between the initial volume of liquid in a graduated cylinder and the volume of liquid with immersed material gives us the volume of the material. That is, the increase in volume after addition of solid sample is equal to the solid volume.

In the liquid displacement method, liquids used should have a low surface tension and should be absorbed very slowly by the particles. Most commonly used fluids are water, alcohol, toluene, and tetrachloroethylene. For displacement, it is better to use a nonwetting fluid such as mercury. Coating of a sample with a film or paint may be required to prevent liquid absorption.

For larger objects, a platform scale can be used (Mohsenin, 1970) (Fig. 1.8). The sample is completely submerged in liquid such that it does not make contact with the sides or bottom of the beaker. Weight of the liquid displaced by the solid sample is divided by its density. The method is based on the Archimedes principle, which states that a body immersed in a fluid will experience a weight loss in an amount equal to the weight of the fluid it displaces. That is, the upward buoyancy force exerted

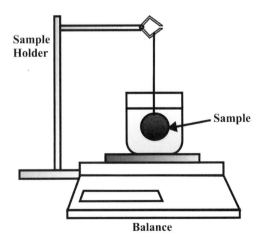

Figure 1.8 Platform scale for measurement of volume of large objects.

on a body immersed in a liquid is equal to the weight of the displaced liquid.

$$
\begin{aligned}
V_s &= \frac{G}{\rho_l} \\
&= \frac{W_{\text{air}} - W_l}{\rho_l}
\end{aligned}
\tag{1.31}
$$

where

G = the buoyancy force (N),
ρ_l = the density of liquid (kg/m^3),
W_{air} = the weight of sample in air (kg),
W_l = the weight of sample in liquid (kg).

Liquids having a density lower than that of sample should be used if partial floating of the sample is observed. The sample is forced into the liquid by means of a sinker rod if it is lighter or it is suspended with a string if it is heavier than the liquid. If the sample is forced into the fluid using a sinker rod, it should be taken into account in the measurement as:

$$
V_s = \frac{G_{\text{sample} + \text{sinker}} - G_{\text{sinker}}}{\rho_l}
\tag{1.32}
$$

1.4.2   Gas Displacement Method

Volumes of particulate solids and materials with irregular shape can be determined by displacement of gas or air in pycnometer (Karathanos & Saravacos, 1993). The most commonly used gases are helium and nitrogen. The pycnometer consists of two airtight chambers of nearly equal volumes, V_1 and V_2, that are connected with small-diameter tubing (Fig. 1.9). The material to be measured is placed in the second chamber. The exhaust valve (valve 3) and the valve between the two chambers (valve 2) are closed. The inlet valve (valve 1) is opened and the gas is supplied to the first chamber until the

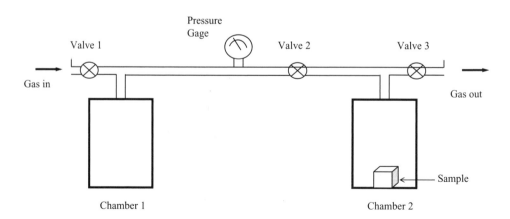

Figure 1.9   Gas comparison pycnometer.

gauge pressure is increased up to a suitable value (e.g., 700–1000 Pa). Then, the inlet valve is closed and the equilibrium pressure is recorded. Assuming that the gas behaves ideally:

$$P_1 V_1 = n R T_1 \tag{1.33}$$

where

P_1 = equilibrium pressure when valve 2 is closed (Pa),
V_1 = volume of the first chamber (m^3),
n = moles of gas (kg mol),
R = gas constant (8314.34 J/kg mol K),
T_1 = absolute temperature (K).

After the equilibrium pressure is recorded, the valve between the two chambers is opened (valve 2) and the gas within the first chamber is allowed to fill the empty spaces (pores) in the second chamber. The new pressure (P_2) is recorded. When valve 2 is opened, total mass of gas (m) is divided into two, one of which fills the first tank (m_1) and the other fills the pore space of the second tank (m_2).

$$m = m_1 + m_2 \tag{1.34}$$

Assuming that the system is isothermal:

$$P_1 V_1 = P_2 V_1 + P_2 V_{a2} \tag{1.35}$$

where V_{a2} is the volume of the empty spaces within the second chamber and can be expressed as:

$$V_{a2} = V_2 - V_s = V_1 \frac{P_1 - P_2}{P_2} \tag{1.36}$$

where V_s is the volume of the solid (m^3) and can be calculated from the following equation:

$$V_s = V_2 - V_1 \left(\frac{P_1 - P_2}{P_2} \right) \tag{1.37}$$

The errors in this method may come from not taking into account the volumes of the tubing connecting the chambers. Moreover, although the calculation assumes an ideal gas, the air does not exactly follow the ideal gas law. In addition, the equalization in pressures between the two chambers is not isothermal. To eliminate these errors, the instrument should be calibrated by using an object of precisely known volume.

1.4.3 Solid Displacement Method

The volume of irregular solids can also be measured by sand, glass bead, or seed displacement method. Rapeseeds are commonly used for determination of volume of baked products such as bread. In the rapeseed method, first the bulk density of rapeseeds is determined by filling a glass container of known volume uniformly with rapeseeds through tapping and smoothing the surface with a ruler. All measurements are done until the constant weight is reached between the consecutive measurements. The densities of the seeds are calculated from the measured weight of the seeds and volume of the container.

Then, the sample and rapeseeds are placed together in the container. The container is tapped and the surface is smoothed with a ruler. Tapping and smoothing are continued until a constant weight is reached between three consecutive measurements. The volume of the sample is calculated as follows:

$$W_{seeds} = W_{total} - W_{sample} - W_{container} \tag{1.38}$$

$$V_{seeds} = \frac{W_{seeds}}{\rho_{seeds}} \tag{1.39}$$

$$V_{sample} = V_{container} - V_{seeds} \tag{1.40}$$

where

W = weight (kg),
V = volume (m^3),
ρ = density (kg/m^3).

1.4.4   Expressions of Volume

Volume can be expressed in different forms. The form of the volume must be well defined before the data are presented. The most commonly used definitions are:

Solid volume (V$_s$) is the volume of the solid material (including water) excluding any interior pores that are filled with air. It can be determined by the gas displacement method in which the gas is capable of penetrating all open pores up to the diameter of the gas molecule.

Apparent volume (V$_{app}$) is the volume of a substance including all pores within the material (internal pores). Apparent volume of regular geometries can be calculated using the characteristic dimensions. Apparent volume of irregularly shaped samples may be determined by solid or liquid displacement methods.

Bulk volume (V$_{bulk}$) is the volume of a material when packed or stacked in bulk. It includes all the pores enclosed within the material (internal pores) and also the void volume outside the boundary of individual particles when stacked in bulk (external pores).

For baked products, especially for cakes, sometimes an index of volume based on the dimensions of the cake is used (Cloke, Davis, & Gordon, 1984). In this method, the cake is cut into two halves. A template is used to measure height from different positions of the cross section (Fig. 1.10). Volume index determined by the AACC template method is based on the sum of height at different positions (AACC, 1983).

$$\text{Volume index} = B + C + D \tag{1.41}$$

The bottom diameter (A to E) was also measured and subtracted from the diameter of the baking pan to obtain shrinkage value. Uniformity, which is a measurement of cake symmetry, is found through subtraction of the two midpoint measurements:

$$\text{Uniformity} = B - D \tag{1.42}$$

1.5   DENSITY

Quality of food materials can be assessed by measuring their densities. Density data of foods are required in separation processes, such as centrifugation and sedimentation and in pneumatic and hydraulic transport of powders and particulates. In addition, measuring the density of liquid is required to determine the power required for pumping.

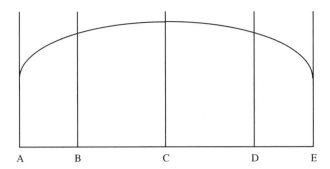

A   B   C   D   E

Figure 1.10 Schematic cross-sectional tracing of a cake where, C is height at center and B and D are heights at three fifths of distance from center to edge.

Density can be calculated after measuring the mass and volume of the object because it is defined as the mass per unit volume. In the SI system, the unit of density is kg/m^3.

In most of the engineering problems, solids and liquids are assumed to be incompressible, that is, the density is hardly affected by moderate changes in temperature and pressure. Gases are compressible and their densities are affected by changes in temperature and pressure. The densities of gases decrease as temperature increases whereas they increase with increase in pressure. Under moderate conditions, most gases obey the ideal gas law. Molecular weight of any gas in kg (1 kg-mole) occupies 22.4 m^3 at 273 K and 1 atm. For example, density of air can be calculated from:

$$\rho_{air} = \frac{29 \, \text{kg/kgmole}}{22.4 \, \text{m}^3/\text{kgmole}} \tag{1.43}$$

The density of liquids can be determined by using a pycnometer. Wide-mouthed bottles can be used for very viscous materials such as tomato paste, batter, or honey.

Liquid density can also be measured by placing a hydrometer in a beaker filled with the liquid sample (Fig. 1.11). The hydrometer has a stem that extends from a tubular shaped bulb. The diameter of the stem is approximately equal to the diameter of thermometer. The bulb may be filled with a dense material to give it an appropriate weight so that the whole hydrometer sinks in the test liquid to such a depth that the upper calibrated and marked stem is partly above the liquid. The depth to which the hydrometer sinks depends on the density of the fluid displaced. The deeper the hydrometer sinks, the lower the density of the solution. The constant weight hydrometer works on the principle that a floating body displaces its own weight of fluid.

The density of liquid is calculated from the ratio of weight of the hydrometer to the volume of the displaced liquid:

$$\rho_1 = \frac{W}{AX + V} \tag{1.44}$$

where

W = weight of hydrometer (kg),
A = cross-sectional area of stem (m^2),
X = the length of the stem immersed (m),
V = Volume of the bulb (m^3).

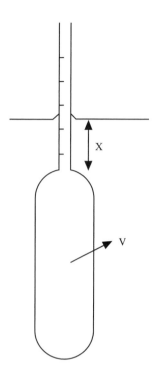

Figure 1.11   Hydrometer.

Density hydrometers are sometimes prepared for a narrow range of measurement and therefore are sensitive to small changes in density. Specific names are given to these kinds of hydrometers such as lactometers for milk and oleometers for oil. The Twaddell hydrometer is used for liquids denser than water. The Baume scale has two scales, one of which is for fluids heavier than water and the other one is for lighter fluids. A variety of hydrometers are also available for specific purposes other than density such as brix saccharometers for percentage of sucrose by weight in a solution, alcoholometers for percentage of alcohol by volume, and salometers for determination of the percent saturation of salt solutions.

The density of solids can be calculated from their measured weight and volume. Volume measurement methods have been discussed in Section 1.4.

Density can be expressed in different forms. For example, for particulate materials such as grains, one may be interested in the density of individual particles or the density of the bulk of the material which includes the void volume. In literature, the definitions of densities differ form each other. Therefore, the form of the density must be well defined before presenting the data. The most commonly used definitions are:

True density (ρ_T) is the density of a pure substance or a composite material calculated from the densities of its components considering conservation of mass and volume. If the densities and volume or mass fractions of constituents are known, density can be determined from:

$$\rho_T = \sum_{i=1}^{n} X_i^v \rho_i = \frac{1}{\sum_{i=1}^{n} X_i^w / \rho_i} \qquad (1.45)$$

where

ρ_i = density of ith component (kg/m^3),
X_i^v = volume fraction of ith component,
X_i^w = mass fraction of ith component,
n = number of components.

Solid density (ρ_s) is the density of the solid material (including water), excluding any interior pores that are filled with air. It can be calculated by dividing the sample weight by solid volume determined by the gas displacement method in which gas is capable of penetrating all open pores up to the diameter of the gas molecule.

Material (substance) density (ρ_m) is the density of a material measured when the material has been broken into pieces small enough to be sure that no closed pores remain.

Particle density (ρ_p) is the density of a particle that has not been structurally modified. It includes the volume of all closed pores but not the externally connected ones. It can be calculated by dividing the sample weight by particle volume determined by a gas pycnometer.

Apparent density (ρ_{app}) is the density of a substance including all pores within the material (internal pores). Apparent density of regular geometries can be determined from the volume calculated using the characteristic dimensions and mass measured. Apparent density of irregularly shaped samples may be determined by solid or liquid displacement methods.

Bulk density (ρ_{bulk}) is the density of a material when packed or stacked in bulk. Bulk density of particulate solids is measured by allowing the sample to pour into a container of known dimensions. Special care should be taken since the method of filling and the container dimensions can affect the measurement. It depends on the solid density, geometry, size, surface properties, and the method of measurement. It can be calculated by dividing the sample weight by bulk volume.

The density of food materials depends on temperature and the temperature dependence of densities of major food components [pure water, carbohydrate (CHO), protein, fat, ash and ice] has been presented by Choi and Okos (1986) as follows:

$$\rho_{water} = 997.18 + 3.1439 \times 10^{-3}T - 3.7574 \times 10^{-3}T^2 \tag{1.46}$$

$$\rho_{CHO} = 1599.1 - 0.31046T \tag{1.47}$$

$$\rho_{protein} = 1330 - 0.5184T \tag{1.48}$$

$$\rho_{fat} = 925.59 - 0.41757T \tag{1.49}$$

$$\rho_{ash} = 2423.8 - 0.28063T \tag{1.50}$$

$$\rho_{ice} = 916.89 - 0.1307T \tag{1.51}$$

where densities (ρ) are in kg/m^3 and temperatures (T) are in °C and varies between −40 and 150°C.

Example 1.4. Calculate the true density of spinach at 20°C having the composition given in Table E.1.4.1.

Solution:

Using the temperature dependent density equations (1.46–1.50), densities of components at 20°C are calculated and given in Table E.1.4.2.

Taking the total mass of the spinach as 100 g, the mass fraction of each component in spinach is found and shown in Table E.1.4.2.

Table E.1.4.1   Composition of Spinach

Component	Composition (%)
Water	91.57
Protein	2.86
Fat	0.35
Carbohydrate	1.72
Ash	3.50

Table E.1.4.2   Density and Mass Fraction (X_i^w) of Components of Spinach

Component	Density (kg/m³)	X_i^w
Water	995.74	0.9157
Protein	1319.63	0.0286
Fat	917.24	0.0035
Carbohydrate	1592.89	0.0172
Ash	2418.19	0.0350

The true density of spinach can be calculated using Eq. (1.45):

$$\rho_T = \frac{1}{\sum\limits_{i=1}^{n} \dfrac{X_i^w}{\rho_i}} \tag{1.45}$$

$$\rho_T = \frac{1}{\dfrac{0.9157}{995.74} + \dfrac{0.0286}{1319.63} + \dfrac{0.0035}{917.24} + \dfrac{0.0172}{1592.89} + \dfrac{0.035}{2418.19}} = 1030.53 \ \frac{\text{kg}}{\text{m}^3}$$

1.6 POROSITY

Porosity is an important physical property characterizing the texture and the quality of dry and intermediate moisture foods. Porosity data is required in modeling and design of various heat and mass transfer processes such as drying, frying, baking, heating, cooling, and extrusion. It is an important parameter in predicting diffusional properties of cellular foods.

Porosity (ε) is defined as the volume fraction of the air or the void fraction in the sample and expressed as:

$$\text{Porosity} = \frac{\text{Void volume}}{\text{Total volume}} \tag{1.52}$$

There are different methods for determination of porosity, which can be summarized as follows:

Figure 1.12 Image of bread sample (scale represents 1 cm).

1. Direct method: In this method, porosity is determined from the difference of bulk volume of a piece of porous material and its volume after destruction of all voids by means of compression. This method can be applied if the material is very soft and no attractive or repulsive force is present between the particles of solid.

2. Optical method: In this method, porosity is determined from the microscopic view of a section of the porous medium. This method is suitable if the porosity is uniform throughout the sample, that is, the sectional porosity represents the porosity of whole sample. Pore size distribution can be determined if a suitable software is used to analyze images.

Image J (http://rsb.info.nih.gov/ij/) is a software used to analyze the pores and to determine area based pore size distribution, median pore diameter, and percent area fraction of pores. This software uses the contrast between the two phases (pores and solid part) in the image (Abramoff, Magelhaes, & Ram, 2004). First, the image is obtained. Then, the scanned color image is converted to gray scale using this software. Using bars of known lengths, pixel values are converted into distance units. Figure 1.12 shows the image of a bread sample. From the image, pore areas are extracted by the software (Fig. 1.13). The porosity based on area fraction for this bread sample is determined to be 0.348. The area-based pore size distributions for the bread are shown in Fig. 1.14.

3. Density method: In this method, porosity is calculated from the measured densities.

Porosity due to the enclosed air space within the particles is named apparent porosity (ε_{app}) and defined as the ratio of total enclosed air space or voids volume to the total volume. It can also be named internal porosity. Apparent porosity is calculated from the measured solid (ρ_s) and apparent density (ρ_{app}) data as:

$$\varepsilon_{app} = 1 - \frac{\rho_{app}}{\rho_s} \tag{1.53}$$

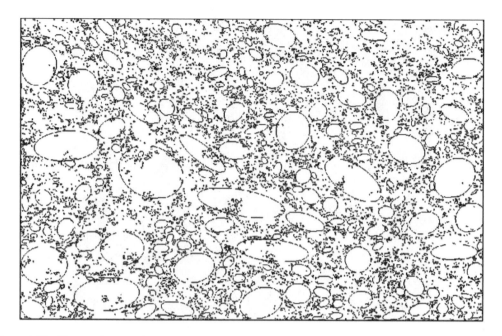

Figure 1.13   Extracted pores of bread using image J.

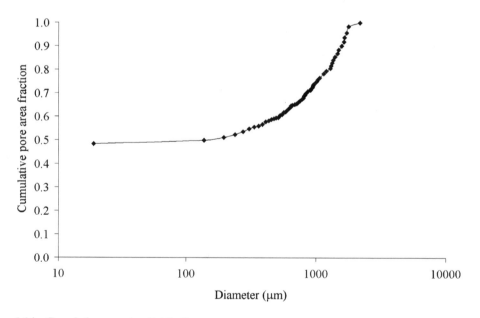

Figure 1.14   Cumulative pore size distribution.

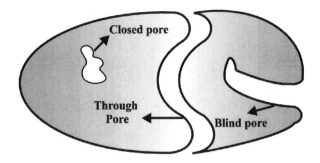

Figure 1.15 Different kinds of pores.

or from the specific solid ($\overline{V}_s$) and apparent ($\overline{V}_{app}$) volumes as:

$$\varepsilon_{app} = 1 - \frac{\overline{V}_s}{\overline{V}_{app}} \tag{1.54}$$

Bulk porosity (ε_{bulk}), which can also be called external or interparticle porosity, includes the void volume outside the boundary of individual particles when stacked as bulk and calculated using bulk and apparent densities as:

$$\varepsilon_{bulk} = 1 - \frac{\rho_{bulk}}{\rho_{app}} \tag{1.55}$$

or from the specific bulk ($\overline{V}_{bulk}$) and apparent ($\overline{V}_{app}$) volumes as:

$$\varepsilon_{bulk} = 1 - \frac{\overline{V}_{app}}{\overline{V}_{bulk}} \tag{1.56}$$

Then, total porosity when material is packed or stacked as bulk is:

$$\varepsilon_{TOT} = \varepsilon_{app} + \varepsilon_{bulk} \tag{1.57}$$

Pores within the food materials (internal pores) can be divided into three groups: closed pores that are closed from all sides, blind pores that have one end closed, and open or flow- through pores where the flow typically takes place (Fig. 1.15).

Since the apparent porosity is due to the enclosed air space within the particles and there are three different forms of pores within the particles, it can be written as:

$$\varepsilon_{app} = \varepsilon_{CP} + \varepsilon_{OP} + \varepsilon_{BP} \tag{1.58}$$

where

ε_{CP} = porosity due to closed pores,
ε_{OP} = porosity due to open or flow through pores,
ε_{BP} = porosity due to blind pores.

Then, total porosity can also be written as:

$$\varepsilon_{TOT} = \varepsilon_{CP} + \varepsilon_{OP} + \varepsilon_{BP} + \varepsilon_{bulk} \tag{1.59}$$

4. Gas pycnometer method: Porosity can be measured directly by measuring the volume fraction of air using the air comparison pycnometer. Remembering Eq. (1.36):

$$V_{a2} = V_2 - V_s = V_1 \frac{P_1 - P_2}{P_2} \tag{1.36}$$

Porosity can be calculated from Eq. (1.36) as:

$$\varepsilon = \frac{V_{a2}}{V_1} = \frac{P_1 - P_2}{P_2} \tag{1.60}$$

5. Using porosimeters: Porosity and pore size distribution can be determined using porosimeters, which are the instruments based on the principle of either liquid intrusion into pores or liquid extrusion from the pores. Pressure is applied to force the liquids, such as water, oil, or mercury, into pores since liquids cannot flow spontaneously into pores. For extrusion porosimetry, wetting liquids are used to fill the pores in the porous materials. Liquid is displaced from the pores by applying differential pressure on the sample and volume of extruded liquid is measured. Pore size distribution of meat patties containing soy protein (Kassama, Ngadi, & Raghavan, 2003), bread and cookie samples (Hicsasmaz & Clayton, 1992), agricultural plant products (Karathanos, Kanellopoulos, & Belessioits, 1996) and starch materials (Karathanos & Saravacos, 1993) have been measured by using a mercury intrusion porosimeter.

In intrusion porosimetry, as intrusion liquid mercury, oil, or water is used. In intrusion porosimetry, liquid is forced into pores under pressure and intrusion volume and pressure are measured. Mercury intrusion porosimetry can measure pores in the size range of 0.03 to 200 μm while nonmercury intrusion porosimetry can measure pores in the size range of 0.001 to 20 μm. This method can detect pore volume, pore diameter, and surface area of through and blind pores. Since very high pressures are required in mercury intrusion, the pore structure of the samples can be distorted.

Extrusion methods can be categorized as capillary flow porosimetry and liquid extrusion porosimetry. Capillary flow porosimetry is a liquid extrusion method in which the differential gas pressure and flow rates through wet and dry samples are measured (Fig. 1.16). A wet sample is put in the chamber. After the chamber is sealed, gas flows into the chamber behind the sample. The point when the pressure can overcome the capillary action of the fluid within the largest pore is recorded as the

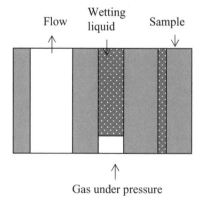

Figure 1.16   Principle of capillary flow porosimetry.

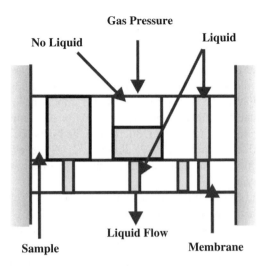

Figure 1.17   Principle of liquid extrusion porosimetry.

---

bubble point. The largest pore should open up at the lowest pressure. For this reason, the pressure at the bubble point is determined and the pore diameter calculated from this pressure is the largest constricted pore. After the bubble point is found, pressure is increased and the flow rate is measured until all the pores are empty. Pore diameters, mean flow pore diameter, through pore surface area, and liquid and gas permeability are determined by this method. Capillary flow porosimetry can measure pore size between 0.013 and 500 μm (Jena & Gupta, 2002).

For large pore sizes, liquid extrusion porosimetry is preferred. Liquid extrusion porosimetry can be used for pore sizes of 0.06 to 1000 μm. A schematic of the liquid extrusion porosimetry is shown in Fig. 1.17. In liquid extrusion porosimetry, the gas pressure applied and the amount of liquid flowing out of a membrane are measured. The pressure required to displace the liquid from a pore is found by equating the work done by the gas to the increase in surface energy, given by:

$$P = \frac{4\sigma \cos\theta}{D} \tag{1.61}$$

where P is the pressure difference across the pore, σ is the surface tension of the wetting liquid, θ is the contact angle of the liquid with the sample, and D is the pore diameter.

As liquid, a low surface tension liquid such as galwick can be used. As shown by Eq. (1.61), lowest pressure will push out the liquid from the largest of the pores, and higher pressures will empty progressively smaller pores. The membrane is chosen such that its pores are smaller than the smallest pores in the sample. Thus, gas pressures required to empty the pores of the sample cannot remove liquid from the pores of the membrane while the liquid pushed out of the sample can pass through the membrane. From the measured pressure and the corresponding volume of liquid collected, the distribution of pore volume as a function of diameter is calculated. The median pore diameter is the diameter at which the volume is equal to half of the total extruded liquid volume.

High mechanical pressure used in pasta extrusion has been found to reduce the porosity and moisture diffusivity of wheat flour (Andrieu & Stamatopoulos, 1986). The formation of pores in foods during drying may show different trends (Rahman, 2003). Porosity may show a maximum or minimum as a function of moisture content. It may also decrease or increase exponentially during drying without showing an optimum point. The porosity of the apple rings increased linearly when moisture content decreased during drying and then reached a constant value (Bai, Rahman, Perera, Smith, & Melton, 2002). A linear increase in the bulk porosity was also observed during drying of the starch samples (Marousis & Saravacos, 1990). The drying method is also important in affecting porosity. Freeze drying was found to produce the highest porosity, whereas in conventional air drying the lowest porosity was observed as compared to vacuum, microwave, and osmotic drying of bananas, apples, carrots, and potatoes (Krokida & Maroulis, 1997). Rahman (2003) developed a theoretical model to predict porosity in foods during drying assuming that volume of pores formed is equal to the volume of water removed during drying.

The presence of pores and degree of porosity affect the mechanical properties of food materials. It has been shown that mechanical properties of extruded food products are affected by porosity (Guraya & Toledo, 1996). Mandala and Sotirakoglou (2005) mentioned that crumb and crust texture of breads could be related to porosity. A third-order regression model was used to express variation of porosity of breads during baking in an infrared–microwave combination oven and the model was found to be significantly dependent on microwave power and halogen lamp power placed at the top of the oven (Demirekler, Sumnu, & Sahin, 2004). Breads baked in an infrared–microwave combination oven had higher porosity than the ones baked in conventional oven.

Porosity is also important in frying, since it affects oil uptake of the product. A linear relationship was found between oil uptake during frying and porosity prior to frying (Pinthus, Weinberg, & Saguy, 1995). Porosity increased during frying of restructured potato product and after a short initial period, it was found to be linearly correlated with oil uptake. An increase in porosity during frying of chicken nuggets was also observed by Altunakar, Sahin, and Sumnu (2004). It was also observed that batter formulation was important in controlling porosity and oil uptake during deep-fat frying. Corn starch was found to provide the highest porosity as compared to amylomaize starch, pregelatinized tapioca starch, and waxy maize starch. Dogan, Sahin, and Sumnu (2005a) showed that porosity of chicken nuggets coated with batter containing rice and soy flour increased during frying but a decrease was observed in the later stages of frying. This may be due to oil intrusion into the pores and capillaries initially filled by air or steam generated from evaporated water. The development of porosity in fried chicken nuggets was found to be a function of protein type used in batter preparation (Dogan, Sahin, & Sumnu, 2005b). Addition of soy protein isolate produced the most porous product, while batters to which whey protein isolate and egg albumin had been added were not significantly different from the control batters. The porosity was modeled as a linear function of moisture loss during frying of meat (Kassama & Ngadi, 2005). An exponential model was used to describe porosity in terms of oil uptake for different frying temperatures. Rahman and Potluri (1990) used a quadratic model to predict porosity as a function of moisture content in calamari mantle meat.

1.7   DETERMINATION OF VOLUME OF DIFFERENT KINDS OF PORES

Total specific pore volume within the material $\overline{V}_{TP}$ can be calculated if the specific volume of all kinds of pores—closed pores ($\overline{V}_{CP}$), blind pores, ($\overline{V}_{BP}$), and flow-through pores ($\overline{V}_{OP}$) are known:

$$\overline{V}_{TP} = \overline{V}_{CP} + \overline{V}_{BP} + \overline{V}_{OP} \tag{1.62}$$

Total specific pore volume within the material can be calculated by measuring specific bulk ($\overline{V}_{\text{bulk}}$) and the specific solid volume determined after compacting the sample to exclude all the pores ($\overline{V}_c$):

$$\overline{V}_{TP} = \overline{V}_{\text{bulk}} - \overline{V}_c \tag{1.63}$$

The difference between the specific solid volume determined by gas pycnometer ($\overline{V}_s$) and specific solid volume after compacting the sample ($\overline{V}_c$), gives the volume of closed pores since in a gas pycnometer, gas enters into the open and blind pores but not the closed ones. From these results, the specific volume of closed pores can be calculated:

$$\overline{V}_{CP} = \overline{V}_s - \overline{V}_c \tag{1.64}$$

Volume of flow through or open pores of the sample, ($\overline{V}_{OP}$), can be measured directly using liquid extrusion porosimetry.

From Eq. (1.62), the specific volume of the blind pores is:

$$\overline{V}_{BP} = \overline{V}_{TP} - \overline{V}_{CP} - \overline{V}_{OP} \tag{1.65}$$

Substituting Eqs. (1.63) and (1.64) into Eq. (1.65):

$$\overline{V}_{BP} = \overline{V}_{\text{bulk}} - \overline{V}_s - \overline{V}_{OP} \tag{1.66}$$

The fraction of open, closed, or blind pores can be calculated by dividing the specified pore volume by total pore volume.

Example 1.5. The porosity of dried apples was measured using an air comparison pycnometer with two identical chambers (Fig. 1.9). The dried apple sample was placed in chamber 2. Valves 2 and 3 were closed and air was supplied to chamber 1. Valve 1 was closed and pressure P_1 was read as 0.508 atm. Then, valve 2 was opened and the new equilibrium pressure, P_2, was read as 0.309 atm. Calculate the porosity of the dried apple.

Solution:

By using Eq. (1.35):

$$P_1 V_1 = P_2 V_1 + P_2 V_{a2} \tag{1.35}$$

$$\varepsilon = \frac{V_{a2}}{V_1} = \frac{P_1 - P_2}{P_2} \tag{1.60}$$

$$\varepsilon = \frac{0.508 - 0.309}{0.309} = 0.644$$

Example 1.6. Cherry has a moisture content of 77.5% (wb). The apparent and bulk densities are 615 kg/m^3 and 511 kg/m^3 at 25°C, respectively. Assuming cherry contains only carbohydrate and water, calculate the total porosity of cherries when stacked as bulk. The densities of carbohydrate and water are 1586 kg/m^3 and 997 kg/m^3, respectively.

Solution:

Since the sample is stacked in bulk, the total porosity should be calculated by considering both enclosed air space within the cherry and void volume between the cherries.

Apparent porosity, which is the ratio of total enclosed air space or void volume to the total volume, can be calculated by using Eq. (1.53):

$$\varepsilon_{app} = 1 - \frac{\rho_{app}}{\rho_s} \tag{1.53}$$

Solid density can be calculated using the true density Eq. (1.45):

$$\rho_T = \frac{1}{\sum\limits_{i=1}^{n} \frac{X_i^w}{\rho_i}} \tag{1.45}$$

$$\rho_T = \frac{1}{\frac{0.775}{997} + \frac{0.225}{1586}} = 1087.9 \frac{kg}{m^3}$$

Then, apparent porosity is:

$$\varepsilon_{app} = 1 - \frac{615}{1087.9} = 0.43$$

Bulk porosity, which includes the void volume outside the individual cherries when stacked as bulk, is:

$$\varepsilon_{bulk} = 1 - \frac{\rho_{bulk}}{\rho_{app}} \tag{1.55}$$

$$\varepsilon_{bulk} = 1 - \frac{511}{615} = 0.17$$

Then, total porosity is:

$$\varepsilon_{TOT} = \varepsilon_{app} + \varepsilon_{bulk} \tag{1.57}$$

$$\varepsilon_{TOT} = 0.43 + 0.17$$

$$= 0.60$$

Example 1.7. Rapeseeds are commonly used for volume measurement of baked goods. In a bakery, first the bulk density of rapeseeds was determined by filling a container of 100 g weight and 1000 cm³ volume uniformly with rapeseeds through tapping and smoothing the surface with a ruler. The weight of the container filled with rapeseeds after constant weight was reached was 750 g.

Then, the volumes of the muffins were measured at different baking times using rapeseeds before and after compressing the pores using a 1000 N load cell and the experimental data are given in Table E.1.7.1. Calculate the porosity of muffins during baking. How does it change during baking? Explain the reason.

Solution:

First, the density of the rapeseeds is calculated as:

$$\rho_{seeds} = \frac{W_{seeds}}{V_{container}}$$

$$= \frac{(750 - 100)\,g}{1000\,cm^3}$$

$$= 0.65\,g/cm^3$$

Table E.1.7.1 Experimental Data Obtained from Rapeseed Method Before and after Compressing the Pores of Muffin Samples During Baking

		Weight (empty container + sample + rapeseeds) (g)	
Baking Time (min)	Weight of Sample (g)	Before Compression	After Compression
10	34	731	744
20	30	718	752
30	28	704	758

Then, the apparent volume of the sample after 10 min of baking is determined using the data before compression and using Eq. (1.40) given below:

$$V_{sample} = V_{container} - V_{seeds} \qquad (1.40)$$

$$= V_{container} - \frac{W_{seeds}}{\rho_{seeds}}$$

$$V_{sample} = 1000\,cm^3 - \frac{(731 - 100 - 34)g}{0.65\,g/cm^3}$$

$$= 81.5\,cm^3$$

Then, solid volume (V_s) of the sample after 10 min of baking is determined by using weight data after compression and Eq. (1.40):

$$V_s = 1000\,cm^3 - \frac{(744 - 100 - 34)g}{0.65\,g/cm^3}$$

$$= 61.5\,cm^3$$

Then, porosity is determined from Eq. (1.54):

$$\varepsilon_{app} = 1 - \frac{\overline{V}_s}{\overline{V}_{app}} \qquad (1.54)$$

$$\varepsilon_{app} = 1 - \frac{61.5}{81.5}$$

$$= 0.25$$

The results for other baking times are shown in Table E.1.7.2. It was observed that porosity of muffins increased as baking time increased because of moisture loss during baking.

1.8 SHRINKAGE

Shrinkage is the decrease in volume of the food during processing such as drying. When moisture is removed from food during drying, there is a pressure imbalance between inside and outside of the food. This generates contracting stresses leading to material shrinkage or collapse (Mayor & Sereno, 2004). Shrinkage affects the diffusion coefficient of the material and therefore has an effect on the

Table E.1.7.2   Porosity of Muffins During Baking

Time (min)	V_{app}	V_s	Porosity
10	81.5	61.5	0.25
20	95.4	43.1	0.55
30	113.8	30.8	0.73

drying rate. Apparent shrinkage is defined as the ratio of the apparent volume at a given moisture content to the initial apparent volume of materials before processing:

$$S_{app} = \frac{V_{app}}{V_{app_0}} \tag{1.67}$$

where

V_{app} = apparent volume at a given moisture content (m^3),
V_{app_0} = Initial apparent volume (m^3).

Shrinkage is also defined as the percent change from the initial apparent volume. Two types of shrinkage are usually observed in food materials. If there is a uniform shrinkage in all dimensions of the material, it is called *isotropic shrinkage*. The nonuniform shrinkage in different dimensions, on the other hand, is called *anisotropic shrinkage*.

Shrinkage during drying was modeled in the literature by using empirical and fundamental models (Mayor & Sereno, 2004). Empirical models correlates shrinkage with moisture content. Linear models can be used if the porosity development during drying is negligible. Mulet, Tarrazo, Garcia-Reverter, and Berna (1997) showed that if porosity increases sharply during the final stage of drying, shrinkage can be explained best by the exponential model. Fundamental models allow the prediction of moisture content and/or change in volume to be obtained without complicated mathematical calculations. Some of the fundamental models including variation of porosity during drying process are given below (Mayor & Sereno, 2004; Perez & Calvelo, 1984; Rahman, Perera, Chen, Driscoll, & Potluri, 1996):

$$\frac{V}{V_0} = \frac{1}{(1 - \varepsilon)} \left[1 + \frac{\rho_0 (X - X_0)}{\rho_w (1 + X_0)} \right] \qquad \text{(Perez \& Calvelo, 1984)} \tag{1.68}$$

where X is moisture content in dry basis, ρ is the density, the subscript 0 denotes the initial condition, and the subscript w denotes water.

The model proposed by Perez and Calvelo (1984) does not need compositional data of the solid phase to calculate shrinkage. This model was improved by taking the initial porosity of the material into account (Mayor & Sereno, 2004).

$$\frac{V}{V_0} = \frac{1}{(1 - \varepsilon)} \left[1 + \frac{\rho_0 (X - X_0)}{\rho_w (1 + X_0)} - \varepsilon_0 \right] \qquad \text{(Mayor \& Sereno, 2004)} \tag{1.69}$$

Most of the fundamental models assume addition of the volumes of different phases in the system. An exception to this volume addition is seen in the model proposed by Rahman et al. (1996). Interaction between the phases of the material is accounted for by means of the excess volume due to the interaction of component phases.

$$\frac{V}{V_0} = \frac{\rho_0}{\rho}\left[\frac{1+X}{1+X_0}\right] \qquad \text{(Rahman et al., 1996)} \qquad (1.70)$$

where X_i^w is mass fraction and $\rho = \dfrac{(1 - \varepsilon_{ex} - \varepsilon)}{\displaystyle\sum_{i=1}^{m}\frac{X_i^w}{(\rho_T)_i}}$

The excess volume fraction (ε_{ex}) is the ratio of the excess volume, which is defined as the change in volume that results from the mixture of the pure components at a given temperature and pressure to the total volume.

PROBLEMS

1.1. Calculate the sphericity of a peach having major, intermediate, and minor diameters of 58.2 mm, 55.2 mm, and 48.8 mm, respectively.

1.2. Determine the porosity of a sweet Vidali onion, which is composed of 93.58% water, 1.04% protein, 0.45% fat, 4.48% carbohydrate, and 0.46% ash. The apparent density of the onion was measured by the liquid displacement method. Toluene was used as the pycnometer liquid because the onion pieces float on water. All the measurements were carried out at 20°C and the density of toluene at that temperature is known to be 865 kg/m^3. The data obtained from liquid displacement measurements are:

Weight of the empty pycnometer:　　　　　　　　　75.87 g
Weight of the pycnometer filled with toluene:　　　126.58 g
Weight of the pycnometer containing only onion:　　85.87 g
Weight of the pycnometer containing onion and　　　127.38 g
filled with toluene:

Calculate the porosity of the sweet Vidalia onion.

1.3. A pycnometer consisting of two chambers of equal volumes (35 cm^3) was used for the determination of bulgur density (Fig. 1.9). Ten grams of bulgur was placed in the second chamber of the pycnometer. Valves 2 and 3 were closed and air was supplied to chamber 1. Valve 1 was closed and pressure P_1 was read as 134 kPa. Then, valve 2 was opened and the new equilibrium pressure P_2 was read as 74.4 kPa. The same bulgur is poured into a graduated cylinder and 100 mL of the bulgur is determined to have a weight of 74 g. Calculate the bulk porosity of the bulgur.

1.4. The effect of 1% (w/w) sucrose ester addition as an emulsifier in white pan bread is examined. As a control, bread prepared without addition of emulsifier was used. Specific volumes of breads were determined by using the rapeseed method. First, the weight of 100-mL tapped rapeseeds is determined to be 76 g. The bread samples of known weight and rapeseeds are then placed together into a container of 2580 g with 2-L inner volume. The container is tapped and the surface is smoothed with a ruler. The total weight of the container and its contents is measured for each sample and given in Table P.1.4.1. Calculate the bread specific volumes and comment on the effect of the emulsifier on the specific volume.

1.5. A sample containing spherical particles has a particle size distribution as given in Table P.1.5.1. Calculate the Sauter mean diameter.

Table P.1.4.1   Data Obtained for Measurement of Bread Volume

Sample	Bread Weight (g)	Total Weight (g)
Control bread	193.9	3612.9
Bread with emulsifier	185.8	3491.6

Table P.1.5.1   Particle Size Distribution of Sample

Percent Weight of Material of Diameter Less Than D	D(mm)
0	0.38
10	0.52
20	0.54
40	0.62
55	0.68
75	0.74
90	0.84
100	0.92

REFERENCES

AACC (1983). *Approved Methods of the AACC.* Method 10–91, approved April 1968, revised October, 1982. St. Paul, MN: The Association.

Abramoff, M.D., Magelhaes, P.J., & Ram, S.J. (2004). Image processing with Image J. *Biophotonics International, 11,* 36–42.

Altunakar, B., Sahin, S., & Sumnu, G. (2004). Functionality of batters containing different starch types for deep-fat frying of chicken nuggets. *European Food Research and Technolology, 218,* 318–322.

Andrieu, J., & Stamatopoulos, A.A. (1986). Moisture and heat transfer modeling during durum wheat drying. In A.S. Mujumdar (Ed.), *Drying* 8 (Vol. 2, p. 492). New York: Hemisphere.

Bai, Y., Rahman, M.S., Perera, C.O., Smith, B., & Melton, L.D. (2002). Structural changes in apple rings during convection air-drying with controlled temperature and humidity. *Journal of Agricultural and Food Chemistry, 50,* 3179–3185.

Bayram, M. (2005). Determination of the sphericity of granular food materials. *Journal of Food Engineering, 68,* 385–390.

Chen, J., Lii, C., & Lu, S. (2004). Relationship between grain physicochemical characteristics and flour particle size distribution for Tawian rice cultivars. *Journal of Food and Drug Analysis, 12,* 52–58.

Choi, Y., & Okos, M.R. (1986). Effects of temperature and composition on the thermal properties of foods. In M. Le Maguer, & P. Jelen (Eds.), *Food Engineering and Process Applications,* Vol. 1: *Transport Phenomena.* New York: Elsevier.

Cloke, J.D., Davis, E.A., & Gordon, J. (1984). Volume measurements calculated by several methods using cross-sectional tracings of cake. *Cereal Chemistry, 61,* 375–377.

Demirekler, P., Sumnu, G., & Sahin, S. (2004). Optimization of bread baking in halogen lamp-microwave combination oven by response surface methodology. *European Food Research and Technology, 219,* 341–347.

Dogan, S.F., Sahin, S., & Sumnu, G. (2005a). Effects of soy and rice flour addition on batter rheology and quality of deep-fat fried chicken nuggets. *Journal of Food Engineering, 71,* 127–132.

Dogan, S.F., Sahin, S., & Sumnu, G. (2005b). Effects of batters containing different protein types on quality of deep-fat fried chicken nuggets. *European Food Research and Technology, 220,* 502–508.

Guraya, H.S., & Toledo, R.T. (1996). Microstructural characteristics and compression resistance as indices if sensory texture in a crunchy snack product. *Journal of Texture Studies, 27,* 687–701.

Hebrard, A., Oulahna, D., Galet, L., Cuq, B., Abecassis, J., & Fages, J. (2003). Hydration properties of durum wheat semolina: Influence of particle size and temperature. *Powder Technology, 130,* 211–218.

Hicsasmaz, Z., & Clayton, J.T. (1992). Characterization of the pore structure of starch based food materials. *Food Structure, 11,* 115–132.

Jena, A., & Gupta, K. (2002). Characterization of pore structure of filtration media. *Fluid Particle Separation Journal, 4,* 227–241.

Karathanos, V.T., & Saravacos, G.D. (1993). Porosity and pore size distribution of starch materials. *Journal of Food Engineering, 18,* 259–280.

Karathanos, V.T., Kanellopoulos, N.K., & Belessiotis, V.G. (1996). Development of porous structure during air drying of agricultural plant products. *Journal of Food Engineering, 29,* 167–183.

Kassama, L.S., & Ngadi, M.O. (2005). Pore development and moisture transfer in chicken meat during deep-fat frying. *Drying Technology, 23,* 907–923.

Kassama, L.S., Ngadi, M.O., & Raghavan, G.S.V. (2003). Structural and instrumental textural properties of meat patties containing soy protein. *International Journal of Food Properties, 6,* 519–529.

Krokida, M.K., & Maroulis, Z.B. (1997). Effect of drying method on shrinkage and porosity. *Drying Technology, 15,* 2441–2458.

Maduako, J.N., & Faborode, M.O. (1990). Some physical properties of cocoa pods in relation to primary processing. *Ife Journal of Technology, 2,* 1–7.

Mandala, I.G., & Sotirakoglou, K. (2005). Effect of frozen storage and microwave reheating on some physical attributes of fresh bread containing hydrocolloids. *Food Hydrocolloids, 19,* 709–719.

Marousis, S.N., & Saravacos, G.D. (1990). Density and porosity in drying starch materials. *Journal of Food Science, 55,* 1367–1372.

Mayor L., & Sereno, A.M. (2004). Modelling shrinkage during convective drying of food material: A review. *Journal of Food Engineering, 61,* 373–386.

McCabe, W.L., & Smith, J.C. (1976). *Unit Operations of Chemical Engineering,* 3rd ed. Singapore: McGraw-Hill.

McCabe, W.L, Smith, J.C., & Harriot, P. (1993). *Unit Operations of Chemical Engineering,* 5th ed. Singapore: McGraw-Hill.

Mohsenin, N.N. (1970). *Physical Properties of Plant and Animal Materials.* New York: Gordon and Breach.

Mulet, A., Tarrazo, J., Garcia-Reverter, J., & Berna A. (1997). Shrinkage of cauliflower florets and stems during drying. In R. Jowitt (Ed.), *Engineering of Food at ICEF 7* (pp. 97–100). Sheffield, UK: Sheffield Academic Press.

Perez, M.G.R., & Calvelo, A. (1984). Modeling the thermal conductivity of cooked meat. *Journal of Food Science, 49,* 152–156.

Pinthus, E.J., Weinberg, P., & Saguy, I.S. (1995). Oil uptake in deep fat frying as affected by porosity, *Journal of Food Science, 60,* 767–769.

Rahman, M.S. (2003). A theoretical model to predict the formation of pores in foods during drying. *International Journal of Food Properties, 6,* 61–72.

Rahman, M.S. (2005). Mass-volume-area-related properties of foods. In M.A. Rao, S.S.H. Rizvi & A.K. Datta (Eds.), *Engineering Properties of Foods,* 3rd ed. (pp. 1–39). Boca Raton, FL: CRC Press Taylor & Francis.

Rahman, M.S., & Potluri, P.L. (1990). Shrinkage and density of squid flesh during air drying. *Journal of Food Engineering, 12,* 133–143.

Rahman, M.S., Perera, C.O., Chen, X.D., Driscoll, R.H., & Potluri, P.L. (1996) Density, shrinkage and porosity of calamari mantle meat during air drying in a cabinet dryer as a function of water content. *Journal of Food Engineering, 30,* 135–145.

Resch, J.J., & Daubert, R.C. (2001). *Institute of Food Technologists Annual Meeting Book of Abstracts,* June 23–27, New Orleans, LA, p. 24.

Sabliov, C.M., Boldor, D., Keener, K.M., & Farkas, B.E. (2002). Image processing method to determine surface area and volume of axi-symmetric agricultural products. *International Journal of Food Properties, 5,* 641–653.

CHAPTER 2

Rheological Properties of Foods

SUMMARY

In this chapter, rheological properties of foods are discussed, concentrating on the principles of flow behavior and deformation of food systems. The principles of viscosity and texture measurement methods and the devices used in these methods are explained in detail. In addition, models used to understand the rheology of food materials are discussed.

Rheological properties are defined as mechanical properties that result in deformation and the flow of material in the presence of a stress. The viscosity is constant and independent of shear rate in Newtonian fluids. If the fluid is non-Newtonian, its viscosity may increase or decrease with increasing shear rate. For shear thinning fluids viscosity decreases with increasing shear rate while for shear thickening fluids viscosity increases with increasing shear rate. A yield stress is required for plastic fluids to flow. For time-dependent fluids, viscosity changes with respect to time. Capillary flow, orifice type, falling ball, and rotational viscometers are the most commonly used viscometers to measure viscosity of materials. Foods showing both elastic and viscous components are known as viscoelastic foods. Viscoelastic materials can be determined by stress relaxation test, creep test, and dynamic test. The Maxwell model is used to interpret stress relaxation of viscoelastic liquids. In the Maxwell model, the spring and the dashpot are connected in series. The Kelvin-Voigt model is used to describe creep behavior that contains a spring and a dashpot connected in parallel. A series combination of Kelvin and Maxwell model is known as the Burger model.

The texture profile of food materials including properties such as hardness, gumminess, adhesiveness, cohesiveness, fracturability, springiness, and chewiness can be determined by using a texture analyzer. Dough rheology can be studied by using a farinograph, mixograph, extensograph, and alveograph.

2.1 INTRODUCTION TO RHEOLOGY

Rheology is the science that studies the deformation of materials including flow. Rheological data are required in product quality evaluation, engineering calculations, and process design. An understanding of flow behavior is necessary to determine the size of the pump and pipe and the energy requirements. The rheological models obtained from the experimental measurements can be useful in design of food engineering processes if used together with momentum, energy, and mass balances. Effects of

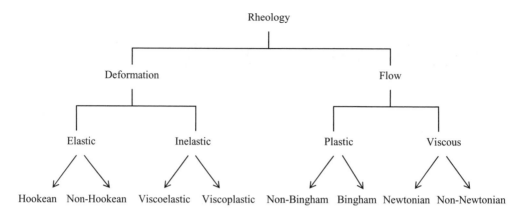

Figure 2.1   Classification of rheology.

processing on rheological properties must be known for process control. Rheology can be classified into different groups as shown in Fig. 2.1.

2.2   FLOW OF MATERIAL

2.2.1   Newton's Law of Viscosity

Consider a fluid between two large parallel plates of area A, separated by a very small distance Y. The system is initially at rest but at time $t = 0$, the lower plate is set in motion in the z-direction at a constant velocity V by applying a force F in the z-direction while the upper plate is kept stationary. At $t = 0$, the velocity is zero everywhere except at the lower plate, which has a velocity V (Fig. 2.2). Then, the velocity distribution starts to develop as a function of time. Finally, steady state is achieved and a linear velocity distribution is obtained. The velocity of the fluid is experimentally found to vary linearly from zero at the upper plate to velocity V at the lower plate, corresponding to no-slip conditions at each plate.

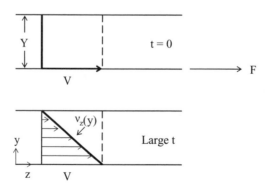

Figure 2.2   Velocity profile in steady flow for a Newtonian fluid between two parallel plates.

Experimental results show that the force required to maintain the motion of the lower plate per unit area is proportional to the velocity gradient, and the proportionality constant, μ, is the viscosity of the fluid:

$$\frac{F}{A} = \mu \frac{V}{Y} \tag{2.1}$$

The microscopic form of this equation is known as Newton's law of viscosity:

$$\tau_{yz} = -\mu \frac{dv_z}{dy} = -\mu \dot{\gamma}_{yz} \tag{2.2}$$

where

$\tau_{yz} =$ shear stress (N/m^2),
$\mu =$ viscosity (Pa·s),
$\dot{\gamma}_{yz} =$ shear rate (1/s).

Shear stress and shear rate have two subscripts: z represents the direction of force and y represents the direction of normal to the surface on which the force is acting. A negative sign is introduced into the equation because the velocity gradient is negative, that is, velocity decreases in the direction of transfer of momentum.

Example 2.1. Two parallel plates are 0.1 m apart. The bottom plate is stationary while the upper one is moving with a velocity V (Fig E.2.2.1). The fluid between the plates is water, which has a viscosity of 1 cp.

(a) Calculate the momentum flux necessary to maintain the top plate in motion at a velocity of 0.30 m/s.
(b) If water is replaced with a fluid of viscosity 100 cp, and momentum flux remains constant, find the new velocity of the top plate.

Solution:

(a) $\mu_w = 1$ cp $= 1 \times 10^{-3}$ Pa · s
   Newton's law of viscosity is used to determine shear stress:

$$\tau_{yx} = -\mu \frac{dv_x}{dy} \tag{2.2}$$

$$\tau_{yx} = -1 \times 10^{-3} \, \text{Pa} \cdot \text{s} \frac{(0 - 0.3) \, \text{m/s}}{(0.1 - 0) \, \text{m}} = 0.003 \, \text{Pa}$$

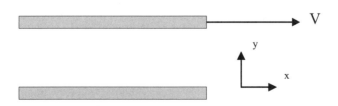

Figure E.2.1.1   Illustration of Example 2.1.

(b) $\mu = 100\,\text{cp} = 0.1\,\text{Pa} \cdot \text{s}$

$$0.003 = -0.1\,\text{Pa} \cdot \text{s}\frac{(0 - V)\,\text{m/s}}{(0.1 - 0)\,\text{m}} \Rightarrow V = 0.003\,\text{m/s}$$

Viscosity is defined as the resistance of a fluid to flow. The unit of dynamic viscosity is (Pa · s) in the SI system and poise (g/cm · s) in the CGS system.

Viscosity varies with temperature. The difference in the effect of temperature on viscosity of liquids and gases is related to the difference in their molecular structure. Viscosity of most of the liquids decreases with increasing temperature.

Theories have been proposed regarding the effect of temperature on viscosity of liquids. According to the Eyring theory, there are vacancies in liquid (Bird, Stewart, & Lightfoot, 1960). Molecules continuously move into these vacancies. This process permits flow but requires energy. This activation energy is more readily available at higher temperatures and the fluid flows easily. The temperature effect on viscosity can be described by an Arrhenius type equation:

$$\mu = \mu_\infty \exp\left(\frac{E_a}{RT}\right) \tag{2.3}$$

where

E_a = activation energy (J/kg mol),
R = gas constant (8314.34 J/kg mol K),
T = absolute temperature (K),
μ_∞ = constant (Pa · s).

Liquid molecules are closely spaced with strong cohesive forces between them. The temperature dependence of viscosity can also be explained by cohesive forces between the molecules (Munson, Young, & Okiishi, 1994). As temperature increases, these cohesive forces between the molecules decrease and flow becomes freer. As a result, viscosities of liquids decrease as temperature increases. In liquids, the intermolecular (cohesive) forces play an important role. Viscosities of liquids show little dependence on density, molecular velocity, or mean free path.

In most liquids, viscosity is constant up to a pressure of 10.134 MPa, but at higher pressures viscosity increases as pressure increases.

In gases, in contrast to liquids, molecules are widely spaced and intermolecular forces are negligible. In most gases, viscosity increases with increasing temperature, which can be expressed by the kinetic theory. The first analysis of viscosity by kinetic theory was made by Maxwell in 1860 (Loeb, 1965). Resistance to relative motion is the result of the exchange of momentum of gas molecules between adjacent layers. As molecules are transported by random motion from a region of low bulk velocity to mix with molecules in a higher velocity region (and vice versa), there is a momentum exchange that resists the relative motion between the layers. As temperature increases the random molecular activity increases, which corresponds to an increase in viscosity. A more detailed discussion on the effect of temperature on viscosities of fluids can be found in Loeb (1965) and Reid, Prausnitz, and Sherwood (1977).

Consider a pure gas composed of rigid, nonattracting spherical molecules of diameter d and mass m present in a concentration of N molecules per unit volume. It is considered that N is small enough so that the average distance between molecules is many times their diameter d. According to kinetic theory, it is assumed that an average molecule traverses a distance equal to the mean free path between impacts. If mean free path is λ, one may consider the length of this path is the thickness of the layer of gas in which viscous action takes place. On the two sides of a gas layer having a thickness of λ, the difference of streaming velocity in the gas is expressed as $\lambda \dfrac{dv}{dz}$, for the velocity gradient normal to the motion of gas, $\dfrac{dv}{dz}$. Molecules coming from upper to the lower layer carry an excess momentum of $m\lambda \dfrac{dv}{dz}$ from the upper to the lower side. It can be said that on the average one third of the molecules are moving with paths that are up or down. Thus, the number of molecules of speed ($\bar{c}$) going up or down per unit area per second will be one third of $N\bar{c}$. The momentum transferred across this layer up and down by the molecules can be expressed as:

$$F = \frac{1}{3}N\bar{c}m\lambda\frac{dv}{dz} \tag{2.4}$$

From Newton's law of viscosity:

$$F = \mu\frac{dv}{dz} \tag{2.5}$$

From Eqs. (2.4) and (2.5):

$$\mu = \frac{1}{3}N\bar{c}m\lambda \tag{2.6}$$

The mean free path is given by the following equation:

$$\lambda = \frac{1}{\sqrt{2}\pi\,d^2 N} \tag{2.7}$$

Inserting Eq. (2.7) into (2.6) gives:

$$\mu = \frac{m\bar{c}}{3\sqrt{2\pi}\,d^2} \tag{2.8}$$

According to kinetic theory, molecular velocities relative to fluid velocity have an average magnitude given by the following equation:

$$\bar{c} = \sqrt{\frac{8RT}{\pi\,N_A m}} \tag{2.9}$$

where N_A is the Avogadro number, m is the mass of the molecule, R is the gas constant, and T is the absolute temperature. Thus,

$$\mu = \frac{2}{3\pi^{3/2}d^2}\sqrt{mKT} \tag{2.10}$$

where K is the Boltzmann constant ($K = R/N_A$).

Equation (2.10) predicts that viscosity should increase with the square root of temperature. Experimental results showed that viscosity increased with temperature more rapidly (Loeb, 1965).

Gases have the lowest viscosity values. Viscosities of gases are constant up to 1 MPa pressure but increase as pressure increases above this level.

Momentum diffusivity or **kinematic viscosity**, which has the same units as thermal diffusivity ($\alpha = k/\rho\,c_p$) in heat transfer and mass diffusivity (D_{AB}) in mass transfer, is defined to make the transport properties analogous. Its unit is m^2/s in SI and stoke (cm^2/s) in CGS. It is the ratio of

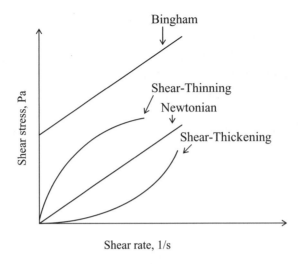

Figure 2.3   The slope of shear stress versus shear rate graph is not constant for non-Newtonian fluids.

---

dynamic viscosity to density of fluid:

$$v = \frac{\mu}{\rho} \tag{2.11}$$

2.2.2 Viscous Fluids

Viscous fluids tend to deform continuously under the effect of an applied stress. They can be categorized as Newtonian or non-Newtonian fluids.

2.2.2.1 Newtonian Fluids

Fluids that follow Newton's law of viscosity (Eq. 2.2) are called Newtonian fluids. The slope of the shear stress versus shear rate graph, which is viscosity, is constant and independent of shear rate in Newtonian fluids (Figs. 2.3 and 2.4). Gases; oils; water; and most liquids that contain more than 90% water such as tea, coffee, beer, carbonated beverages, fruit juices, and milk show Newtonian behavior.

2.2.2.2 Non-Newtonian Fluids

Fluids that do not follow Newton's law of viscosity are known as non-Newtonian fluids. Shear thinning or shear thickening fluids obey the power law model (Ostwald-de Waele equation):

$$\tau_{yz} = k \left(\frac{dv_z}{dy} \right)^n = k(\dot{\gamma}_{yz})^n \tag{2.12}$$

where

$k = $ the consistency coefficient (Pa $\cdot$ s^n),
$n = $ flow behavior index.

Apparent viscosity, Pa·s

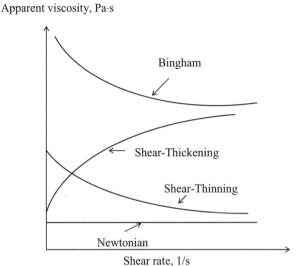

Figure 2.4   Apparent viscosities of time-independent fluids.

For shear thinning (pseudoplastic) fluids $n < 1$,
For shear thickening fluids $n > 1$.

Newtonian fluids can be considered as a special case of this model in which $n = 1$ and $k = \mu$.

The slope of shear stress versus shear rate graph is not constant for non-Newtonian fluids (Fig. 2.3). For different shear rates, different viscosities are observed. Therefore, **apparent viscosity** or a **consistency** term is used for non-Newtonian fluids. The variation of apparent viscosities with shear rates for different types of non-Newtonian fluids is shown in Fig. 2.4.

The symbol η is often used to represent the apparent viscosity to distinguish it from a purely Newtonian viscosity, μ. The ratio of shear stress to the corresponding shear rate is therefore called apparent viscosity at that shear rate:

$$\eta(\dot{\gamma}) = \frac{\tau}{\dot{\gamma}} \tag{2.13}$$

The apparent viscosity and the Newtonian viscosity are identical for Newtonian fluids but apparent viscosity for a power law fluid is:

$$\eta(\dot{\gamma}) = \frac{k(\dot{\gamma})^n}{\dot{\gamma}} = k(\dot{\gamma})^{n-1} \tag{2.14}$$

(a)   Shear Thinning (Pseudoplastic) Fluids. In these types of fluids, as shear rate increases friction between layers decreases. Shearing causes entangled, long-chain molecules to straighten out and become aligned with the flow, reducing viscosity. A typical example for shear thinning fluids is paint. When paint is on the surface but brushing is not applied, its viscosity increases and prevents it from flowing under the action of gravity. When paint is applied to a surface by brushing which shears the paint, its viscosity decreases. Another example for a pseudoplastic fluid is the ink in a ballpoint pen. When the pen is not in use, the ink is so viscous that it does not flow. When we begin to write, the small ball on its point rolls and the turning of ball creates the shearing movement. As a result, the viscosity of ink decreases and it flows on the paper. Fruit and vegetable products such as applesauce, banana puree, and concentrated fruit juices are good examples of pseudoplastic fluids in food systems.

Krokida, Maroulis, and Saravacos (2001) recently analyzed rheological properties of fruit and vegetable products. The consistency coefficient k increased exponentially while the flow behavior index n decreased slightly with concentration. Flow behavior index was close to 0.5 for pulpy products and close to 1.0 for clear juices. While the flow behavior index was assumed to be relatively constant with temperature, the effect of temperature on both apparent viscosity, η and consistency coefficient of the power law model, k was explained by an Arrhenius-type equation.

The flow properties of peach pulp–granular activated carbon mixtures at temperatures of 15 to 40°C and at granular activated carbon concentrations of 0.5 to 5.0 kg/m^3 exhibited a shear thinning behavior (Arslanoglu, Kar, & Arslan, 2005). The flow behavior index and the consistency coefficient of peach pulp–granular activated carbon mixtures were in the ranges of 0.328 to 0.512 and 2.17 to 6.18 Pa · s^n, respectively. Both the consistency coefficient and the flow behavior index decreased with increasing temperature.

Rheological behavior of foods may change depending on concentration. The rheological behavior of concentrated grape juice with a Brix value of 82.1 showed shear thinning behavior (Kaya & Belibagli, 2002). However, diluted samples with a Brix value of 52.1 to 72.9 were found to be Newtonian.

The rheological behavior of sesame paste–concentrated grape juice blends, which is a traditional food product in Turkish breakfasts, was studied at 35 to 65°C and at 20% to 32% sesame paste concentration. All the blends showed shear thinning behavior with a flow behavior index of 0.70 to 0.85 (Arslan, Yener, & Esin, 2005). The consistency coefficient was described by an Arrhenius-type equation.

Sakiyan, Sumnu, Sahin, and Bayram (2004) studied the rheological behavior of cake batter with different fat concentrations and emulsifier types and found that cake batter exhibited shear thinning behavior. The increase in fat content and addition of emulsifier caused a decrease in the apparent viscosity. Flow behavior index was found to be independent of composition of cake batter.

(b) Shear Thickening Fluids. In these types of fluids, as shear rate increases, the internal friction and apparent viscosity increase. A person falling in a swamp tries to escape as soon as possible. However, as he tries to move in panic, sudden shearing is created and the more he tries to escape, the greater the force is required for his movement. Walking on wet sand on a beach is another example of shear thickening fluids. If a sand–water suspension has settled for some time, the void fraction occupied by water is minimal. Any shear will disturb close packing and the void fraction will increase. Water will no longer fill the space between the sand granules and the lack of lubrication will cause an increased resistance to flow.

In food systems, corn starch suspension is an example of shear thickening fluids. Dintiz, Berhow, Bagley, Wu, and Felker (1996) showed the shear thickening phenomenon in unmodified starches (waxymaize, waxyrice, waxybarley, waxypotato, wheat, rice, maize) that had been dissolved and dispersed at 3.0% concentrations in 0.2 N NaOH. Waxy starches (maize, rice, barley, and potato) showed this behavior to a greater extent than did normal wheat, rice, or maize starches. The amylopectin component was responsible for shear thickening properties.

If the increase in viscosity is accompanied by volume expansion, shear thickening fluids are called dilatant fluids. All the dilatant fluids are shear thickening but not all shear thickening fluids are dilatant.

2.2.3 Plastic Fluids

2.2.3.1 Bingham Plastic Fluids

In these types of fluids, fluid remains rigid when the magnitude of shear stress is smaller than the yield stress (τ_0) but flows like a Newtonian fluid when the shear stress exceeds τ_0. Toothpaste is a typical example of Bingham plastic fluid. It does not flow unless the tube is squeezed. In food systems, mayonnaise, tomato paste, and ketchup are examples of this type of fluid. Equation (2.15) shows the behavior of Bingham plastic fluids.

$$\tau_{yz} = \tau_0 + k\left(\frac{dv_z}{dy}\right) \tag{2.15}$$

The apparent viscosities for Bingham plastic fluids can be determined by taking the ratio of shear stress to the corresponding shear rate:

$$\eta(\dot{\gamma}) = \frac{\tau_0 + k(\dot{\gamma})}{\dot{\gamma}} = \frac{\tau_0}{\dot{\gamma}} + k \tag{2.16}$$

2.2.3.2 Non-Bingham Plastic Fluids

In these types of fluids, a minimum shear stress known as yield stress must be exceeded before flow begins, as in the case of Bingham plastic fluids. However, the graph of shear stress versus shear rate is not linear. Fluids of this type are either shear thinning or shear thickening with yield stress.

Fluids that obey the Herschel-Bulkley model (Bourne, 1982) are characterized by the presence of a yield stress term (τ_0) in the power law equation:

$$\tau_{yz} = \tau_0 + k\left(\dot{\gamma}_{yz}\right)^n \tag{2.17}$$

Minced fish paste and raisin paste obey Herschel-Bulkley model. Flow behavior of rice flour-based batter used in fried products was found to obey the Herschel-Bulkley model (Mukprasirt, Herald, & Flores, 2000).

The Casson model (Casson, 1959) is expressed as:

$$(\tau_{yz})^{0.5} = (\tau_0)^{0.5} + k\left(\dot{\gamma}_{yz}\right)^{0.5} \tag{2.18}$$

Molten milk chocolate obeys the Casson model. When the effect of particle size distribution of nonfat solids on the flow characteristic of molten milk chocolate was investigated, Casson yield stress value was correlated with diameter and specific surface area of non fat solids (Mongia & Ziegler, 2000).

2.2.4 Time Dependency

When some fluids are subjected to a constant shear rate, they become thinner (or thicker) with time (Fig. 2.5).

Fluids that exhibit decreasing shear stress and apparent viscosity with respect to time at a fixed shear rate are called thixotropic fluids (shear thinning with time). This phenomenon is probably due to the breakdown in the structure of the material as shearing continues. Gelatin, egg white, and shortening can be given as examples of this type of fluid.

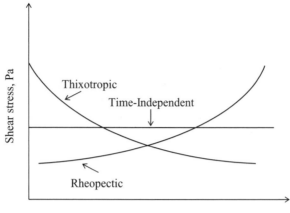

Figure 2.5   Time-dependent behavior of fluids.

Thixotropic behavior may be reversible, partially reversible, or irreversible when the applied shear is removed (fluid is allowed to be at rest). Irreversible thixotropy is called rheomalaxis or rheodestruction (Fig. 2.6).

Thixotropic behavior of a product may be studied by increasing shear stress or shear rate followed by a decrease. If the shear stress is measured as a function of shear rate, as the shear rate is first increased and then decreased, a hysteresis loop will be observed in the shear stress versus shear rate curve (Fig. 2.7).

In rheopectic fluids (shear thickening with time), shear stress and apparent viscosity increase with time, that is, the structure builds up as shearing continues (Fig. 2.5). Bentonite–clay suspensions show this type of flow behavior. It is rarely observed in food systems.

Starch–milk–sugar pastes showed a time dependent flow behavior (Abu-Jdayil & Mohameed, 2004). If the pasting process was done at 85 and 95°C, starch–milk–sugar pastes exhibited a thixotropic

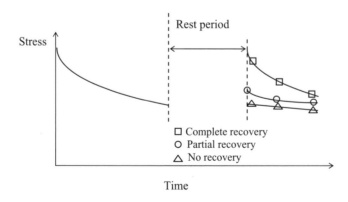

Figure 2.6   Thixotropic behavior observed in torque decay curves.

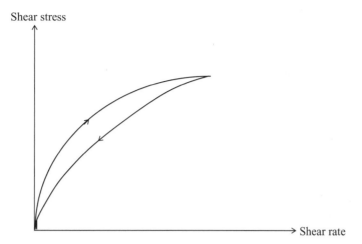

Shear stress

Shear rate

Figure 2.7   Shear stress versus shear rate curve showing hysteresis.

behavior, while pastes processed at 75°C behaved like a rheopectic fluid. It was noted that the thixotropy occurred at high shear stress (above 50 Pa), and the rheopexy occurred at low shear stress (below 45 Pa).

When soy protein was added to tomato juices, thixotropic behavior was observed at low shear rate but this was followed by a transition to rheopectic behavior at higher shear rates (Tiziani & Vodovotz, 2005).

Isikli and Karababa (2005) recently showed that fenugreek paste, which is a local food in Turkey, exhibited rheopectic behavior.

2.2.5 Solution Viscosity

In the case of solutions, emulsions, or suspensions, viscosity is often measured in comparative terms, that is, the viscosity of the solution, emulsion, or suspension is compared with the viscosity of a pure solvent. Solution viscosities are useful in understanding the behavior of some biopolymers including aqueous solutions of locust bean gum, guar gum, and carboxymethyl cellulose (Rao, 1986).

Viscosities of pure solvents and suspensions can be measured and various values can be calculated from the resulting data. The relative viscosity, η_{rel} is expressed as:

$$\eta_{rel} = \frac{\eta_{\text{suspension}}}{\eta_{\text{solvent}}} = 1 + k X_d^v \qquad (2.19)$$

where

X_d^v = volume fraction occupied by the dispersed phase,
k = constant.

The specific viscosity, η_{sp} is:

$$\eta_{sp} = \eta_{rel} - 1 \qquad (2.20)$$

The reduced viscosity, η_{red} is:

$$\eta_{red} = \frac{\eta_{sp}}{C} \tag{2.21}$$

where C is the mass concentration of the solution in g/100 mL.

Inherent viscosity, η_{inh} is:

$$\eta_{inh} = \frac{\ln \eta_{rel}}{C} \tag{2.22}$$

Intrinsic viscosity, η_{int}, can be determined from dilute solution viscosity data:

$$\eta_{int} = \lim_{C \to 0} \left(\frac{\eta_{sp}}{C} \right) \tag{2.23}$$

In dilute solutions, polymer chains are separate and intrinsic viscosity of a polymer in solution depends only on dimensions of the polymer chain (Rao, 1999). The intrinsic viscosities of various protein solutions have been summarized by Rha and Pradipasera (1986).

The equations commonly used for determination of intrinsic viscosity of food gums are Huggins (2.24) and Kramer (2.25) equations (Rao, 1999):

$$\frac{\eta_{sp}}{C} = \eta_{int} + k_1 \eta_{int}^2 C \tag{2.24}$$

$$\frac{\ln \eta_{rel}}{C} = \eta_{int} + k_2 \eta_{int}^2 C \tag{2.25}$$

where k_1 and k_2 are the Huggins and Kramer constants, respectively and they are theoretically related as:

$$k_1 = k_2 + 0.5 \tag{2.26}$$

2.3   VISCOSITY MEASUREMENT

The most commonly used viscosity measurement devices are capillary flow viscometers, orifice type viscometers, falling ball viscometers, and rotational viscometers.

2.3.1   Capillary Flow Viscometers

Capillary flow viscometers are generally in the form of a U-tube. These types of viscometers are very simple, inexpensive, and suitable for low-viscosity fluids. There are different designs of capillary viscometers. A typical design of capillary viscometer is shown in Fig. 2.8.

In capillary flow viscometers, the time for a standard volume of fluid to pass through a known length of capillary tubing is measured. The flow rate of material due to a known pressure gradient is determined. The driving pressure is usually generated by the force of gravity acting on a column of the liquid although it can be generated by the application of compressed air or by mechanical means. Gravity-operated glass capillaries are suitable only for Newtonian fluids having viscosities in the range of 0.4 to 20,000 mPa·s (Steffe, 1996). To measure the viscosities of more viscous fluids, external pressure may be applied. For non-Newtonian fluids, this device is less suitable because the measurement cannot be done at a constant shear rate. Capillary viscometers can be used only for non-Newtonian fluids if the applied external pressure is more significant than static pressure.

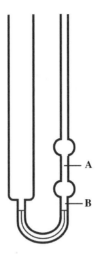

Figure 2.8 Cannon-Fenske capillary flow viscometer.

The diameter of a capillary viscometer should be small enough to provide laminar flow. Capillary viscometers are calibrated with Newtonian oils of known viscosities since the flow rate depends on the capillary radius, which is difficult to measure.

For the viscosity measurement, the viscometer is accurately filled with an accurately known volume of test fluid and the apparatus is immersed in a constant temperature bath until equilibrium is reached. Then, fluid is sucked up from the other limb through the capillary tube until it is above the marked level (A) (Fig. 2.8). Then, suction is removed and fluid flows through the capillary tube under the influence of gravity or the induced pressure head and the time for the fluid to flow from mark A to B is recorded. This time is a direct measure of the kinematic viscosity since it depends on both viscosity and density of fluid. This can be written as:

$$v = Ct \tag{2.27}$$

where C is the calibration constant.

Assuming that the flow is laminar, fluid is incompressible, velocity of the fluid at the wall is zero (no-slip condition), and end effects are negligible, making a macroscopic force balance for a fluid flowing through a horizontal cylindrical tube of length (L) and inner radius (r), the following equation is obtained:

$$\Delta P \pi r^2 = \tau 2\pi r L \tag{2.28}$$

where ΔP is the pressure drop causing flow and τ is the shear stress resisting flow. This equation can be solved for shear stress:

$$\tau = \frac{\Delta P r}{2L} \tag{2.29}$$

For a Newtonian fluid, both shear stress and shear rate vary linearly from zero at the center ($r = 0$) of the capillary to a maximum at the wall ($r = R$). For a Newtonian fluid, this results in the parabolic velocity profile. Then, the shear stress on the fluid at the wall (τ_w) is related to the pressure drop along the length of the tube:

$$\tau_w = \frac{\Delta PR}{2L} \tag{2.30}$$

The flow in capillary viscometers is described by the Hagen Poiseuille equation:

$$\Delta P = \frac{8\mu vL}{R^2} \tag{2.31}$$

Substituting Eq. (2.31) into Eq. (2.30), shear stress can also be expressed as:

$$\tau_w = \mu \frac{4v}{R} \tag{2.32}$$

Then, shear rate at the wall ($\dot{\gamma}_w$) for a Newtonian fluid is given by:

$$\dot{\gamma}_w = \frac{4v}{R} = \frac{4Q}{\pi R^3} \tag{2.33}$$

where Q is the volumetric flow rate. Newton's law of viscosity can be written in terms of pressure gradient and volumetric flow rate as:

$$\frac{\Delta PR}{2L} = \mu \left(\frac{4Q}{\pi R^3} \right) \tag{2.34}$$

and viscosity of the fluid can be determined from the pressure drop and volumetric flow rate or velocity data.

For non-Newtonian fluids, the relation between shear stress and shear rate has to be known to derive these equations. Compared to the parabolic profile for a Newtonian fluid, the profile for a shear thinning fluid is more blunted. The shear rate at the wall can be determined from the Rabinowitsch-Mooney equation (Steffe, 1996; Wilkes, 1999):

$$\dot{\gamma}_w = \left(\frac{3Q}{\pi R^3} \right) + \tau_w \left[\frac{d\left(Q/\pi R^3\right)}{d\tau_w} \right] \tag{2.35}$$

This equation can also be expressed in terms of the apparent wall shear rate, $\dot{\gamma}_{app} = 4Q/\pi R^3$:

$$\dot{\gamma}_w = \left(\frac{3}{4} \right) \dot{\gamma}_{app} + \left(\frac{\tau_w}{4} \right) \left(\frac{d\dot{\gamma}_{app}}{d\tau_w} \right) \tag{2.36}$$

Equation (2.36) can also be written as:

$$\dot{\gamma}_w = \left[\left(\frac{3}{4} \right) + \left(\frac{1}{4} \right) \left(\frac{d(\ln \dot{\gamma}_{app})}{d(\ln \tau_w)} \right) \right] \dot{\gamma}_{app} \tag{2.37}$$

Equation (2.37) can be written in the following simplified form:

$$\dot{\gamma}_w = \left(\frac{3n' + 1}{4n'} \right) \dot{\gamma}_{app} \tag{2.38}$$

where n' is the point slope of the ln (τ_w) versus $\ln(\dot{\gamma}_{app})$. That is:

$$n' = \frac{d(\ln \tau_w)}{d(\ln \dot{\gamma}_{app})} \tag{2.39}$$

If the fluid behaves as a power law fluid, the slope of the derivative is a straight line and $n' = n$.

Table E.2.2.1 Pressure Drop Versus
Volumetric Flow Rate Data for Chocolate
Melt in Capillary Viscometer

Pressure Drop (Pa)	Flow Rate (cm³/s)
3840	0.01
4646	0.06
5762	0.13
6742	0.24
7798	0.37
10,454	0.72
11,760	0.94

Different shear rates can be achieved by varying the flow rate for a single capillary section or by using several capillary sections of different diameters in series.

Example 2.2. The pressure drop versus volumetric flow rate data is obtained for chocolate melt using a capillary viscometer with a pipe diameter of 1 cm and a length of 60 cm (Table E.2.2.1).

(a) Show that chocolate melt is not a Newtonian fluid.
(b) Determine the rheological model constants of the power law, Herschel-Bulkley, and Casson models for the given data.
(c) Which model best represents the rheological behavior of the chocolate melt?

Solution:

(a) Newtonian fluids follow Newton's law of viscosity (Eq. 2.2):

$$\tau_{yz} = -\mu \frac{dv_z}{dy} \tag{2.2}$$

Using pressure drop data, the shear stress at the wall is calculated from Eq. (2.30):

$$\tau_w = \frac{\Delta P R}{2L} \tag{2.30}$$

The shear rate is calculated from Eq. (2.33) for different flow rates:

$$\dot{\gamma}_w = \frac{4Q}{\pi R^3} \tag{2.33}$$

The plot of shear stress versus shear rate is shown in Fig. E.2.2.1. For a fluid to be Newtonian, variation of shear stress versus shear rate should be linear and the intercept must be zero. Since this is not the case, it can be concluded that chocolate melt is not a Newtonian fluid.

(b) The power law equation is:

$$\tau_{yz} = k\left(\frac{dv_z}{dy}\right)^n = k(\dot{\gamma}_{yz})^n \tag{2.12}$$

The power law equation can be linearized by taking the natural logarithm of both sides to determine the flow behavior index (n) and the consistency coefficient (k).

$$\ln \tau_{yz} = \ln k + n \ln \dot{\gamma}_{yz}$$

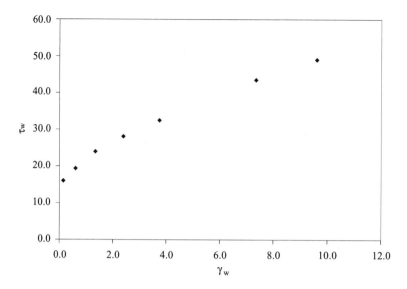

Figure E.2.2.1   Shear stress versus shear rate plot for a chocolate melt.

A logarithmic plot of shear stress versus shear rate ($\ln \tau_{yz}$ versus $\ln \dot{\gamma}_{yz}$) yields a straight line with a slope of n and intercept of $\ln k$ (Fig. E.2.2.2).

$$\ln \tau_w = 3.173 + 0.273 \ln \dot{\gamma}_w$$

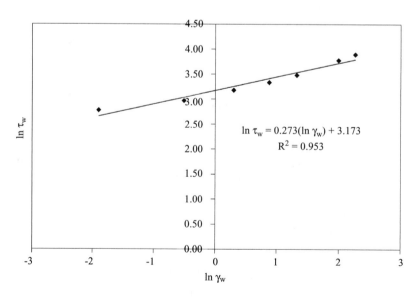

Figure E.2.2.2   The plot of $\ln\tau_w$ versus $\ln\dot{\gamma}_w$ for a chocolate melt.

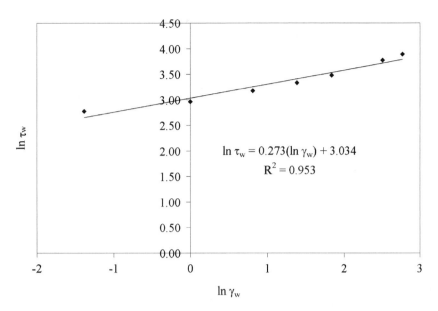

Figure E.2.2.3 Power law model for a chocolate melt.

Being the flow behavior index n to be different than 1, also show that the fluid is not Newtonian. For a power law fluid: $n' = n$

$$\dot{\gamma}_w = \left(\frac{3n' + 1}{4n'}\right) \dot{\gamma}_{app} \tag{2.38}$$

$$\dot{\gamma}_{app} = \frac{4Q}{\pi R^3}$$

Substituting n' and $\dot{\gamma}_{app}$ into Eq. (2.38), $\dot{\gamma}_w$ values are calculated. Then, a logarithmic plot of shear stress versus shear rate ($\ln \tau_w$ versus $\ln \dot{\gamma}_w$) is drawn again (Fig. E.2.2.3).

This plot yields a straight line with a model equation of:

$$\ln \tau_w = 0.273 \ln \dot{\gamma}_w + 3.034$$

The slope of the model gives n, which is 0.273.

From the intercept, which is $\ln k = 3.034$, the value of k is calculated as 20.78 Pa $\cdot$ s^n. The coefficient of determination (r^2) for the model is 0.953.

Thus, the power law expression is:

$$\tau_w = 20.78 \, \dot{\gamma}_w^{0.273}$$

The Herschel-Bulkley expression is given in Eq. (2.17):

$$\tau_w = \tau_0 + k \, (\dot{\gamma}_w)^n \tag{2.17}$$

To determine the value for τ_0, shear stress at the wall (τ_w) is plotted with respect to shear rate at the wall ($\dot{\gamma}_w$) and τ_0 is found to be 13 Pa by extrapolation.

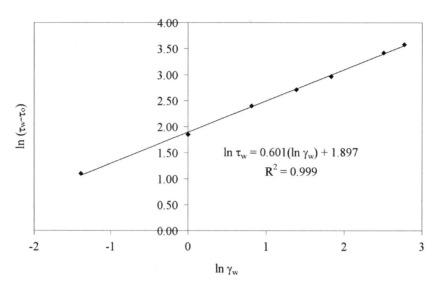

Figure E.2.2.4   Herschel Bulkley model for a chocolate melt.

To find the model constants Herschel-Bulkley expression is linearized as:

$$\ln(\tau_w - \tau_0) = \ln k + n \ln \dot{\gamma}_w$$

The plot of $\ln(\tau_w - \tau_0)$ versus $\ln \dot{\gamma}_w$ is shown in Fig. E.2.2.4.

When linear regression is used, the equation of $\ln(\tau_w - \tau_0) = 1.897 + 0.601 \ln \dot{\gamma}_w$ is determined with a coefficient of determination (r^2) of 0.999.

From the intercept, which is $\ln k = 1.897$, the value of k is calculated as 6.66 Pa · s^n.

Thus, the Herschel-Bulkley expression is:

$$\tau_w = 13 + 6.66 \dot{\gamma}_w^{0.601}$$

The Casson model is given in Eq. (2.18):

$$(\tau_w)^{0.5} = (\tau_0)^{0.5} + k(\dot{\gamma}_w)^{0.5} \tag{2.18}$$

The plot of $(\tau_w)^{0.5}$ versus $(\dot{\gamma}_w)^{0.5}$ is given in Fig. E.2.2.5 and linear regression was performed.

As a result, $\tau_w^{0.5} = 3.570 + 0.861\dot{\gamma}_w^{0.5}$ ($r^2 = 1.000$)

Then, yield stress is;

$$\tau_0 = (3.570)^{1/0.5} = 12.74 \text{ Pa}$$

Thus, the Casson expression is:

$$\tau_w^{0.5} = 12.74^{0.5} + 0.861\dot{\gamma}_w^{0.5}$$

(c) The Casson model is the best model defining the flow behavior of chocolate melt since it has the highest coefficient of determination ($r^2 = 1.000$) compared to others.

Example 2.3. Viscosity of refined sunflower oil was measured at different temperatures by a glass capillary viscometer. Table E.2.3.1 shows the density values and the timing results at different temperatures of sunflower oil. As the reference liquid for calibration of the viscometer, a 50%

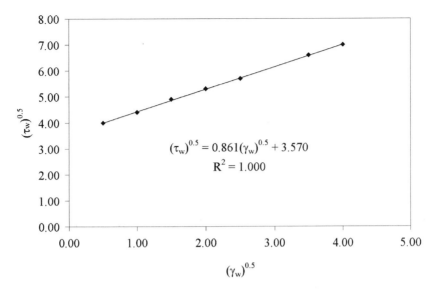

Figure E.2.2.5 Casson model for a chocolate melt.

sucrose solution was used. The density and the viscosity of the reference liquid are known to be 1227.4 kg/m^3 and 0.0126 Pa · s, respectively at 25°C. It took the reference liquid 100 s to fall from one mark to the other of capillary viscometer. Show that the temperature effect on the viscosity of sunflower oil can be expressed by an Arrhenius-type equation. Determine the activation energy and Arrhenius equation constant.

Solution:

Kinematic viscosity is correlated with time as shown in Eq. (2.27):

$$\nu = Ct \tag{2.27}$$

where C is the calibration constant.

Then, the following equation can be written:

$$\frac{\mu}{\mu_{ref}} = \frac{\rho \cdot t}{\rho_{ref} t_{ref}}$$

Table E.2.3.1 Density of Sunflower Oil and Timing Results in Capillary Viscometer at Different Temperatures

Temperature (°C)	Time (s)	Density (kg/m^3)
25	521	916
35	361	899
45	262	883
55	198	867

Table E.2.3.2 Viscosity of
Sunflower Oil at Different
Temperatures

Temperature (°C)	μ (Pa·s)
25	0.049
35	0.033
45	0.024
55	0.018

where μ_{ref} and ρ_{ref} are the viscosity and density of reference liquid, respectively. Inserting the data given in the question for reference fluid, viscosities of sunflower oil at different temperatures are calculated (Table E.2.3.2).

The Arrhenius type equation is:

$$\mu = \mu_\infty \exp\left(\frac{E_a}{RT}\right) \tag{2.3}$$

Taking the natural logarithm of both sides:

$$\ln \mu = \ln \mu_\infty + \left(\frac{E_a}{RT}\right)$$

The plot of $(\ln \mu)$ versus $(1/T)$ is shown in Fig. E.2.3.1.

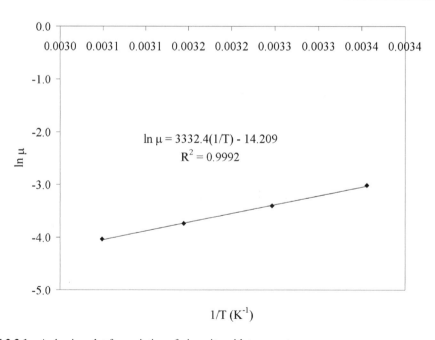

Figure E.2.3.1   Arrhenius plot for variation of viscosity with temperature.

The model equation is found as:

$$\ln \mu = -14.209 + \frac{3332.4}{T}$$

From the intercept of this equation, the Arrhenius constant is found to be 6.75×10^{-7} Ns/m^2. The slope of the equation E_a/R is 3332.4. The gas constant, R, is 8.314×10^{-3} kJ/mole $\cdot$ K. Thus, activation energy, E_a, is calculated as 27.705 kJ/mole.

2.3.2 Orifice Type Viscometers

In orifice type viscometers, the time for a standard volume of fluid to flow through an orifice is measured. They are used for Newtonian or near-Newtonian fluids when extreme accuracy is not required. In the food industry, the most commonly used one is a dipping type Zahn viscometer that consists of a 44-mL capacity stainless steel cup with a handle and with a calibrated circular hole in the bottom. The cup is filled by dipping it into the fluid and withdrawing it. The time from the start of withdrawing to the first break occurring in the issuing stream is recorded.

2.3.3 Falling Ball Viscometers

These types of viscometers involve a vertical tube where a ball is allowed to fall under the influence of gravity. It operates on the principle of measuring the time for a ball to fall through a liquid under the influence of gravity.

When the ball falls through the fluid, it is subjected to gravitational force, drag force, and buoyancy force (Fig. 2.9). Making a force balance:

Net force (F_{Net}) = Gravitational force (F_G) − Buoyancy force (F_B) − Drag force (F_D)

$$\frac{\pi D_p^3 \rho_p}{6} \frac{dv}{dt} = \frac{\pi D_p^3 \rho_p g}{6} - \frac{\pi D_p^3 \rho_f g}{6} - \frac{c_D \pi D_p^2 \rho_f v^2}{8} \tag{2.40}$$

where

D_p = diameter of the ball (m),
ρ_p = density of the ball (kg/m^3),
ρ_f = density of the fluid (kg/m^3),
c_D = drag coefficient,
v = velocity of the ball (m/s).

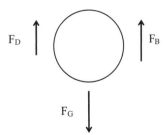

Figure 2.9 Forces acting on a ball in falling ball viscometer.

When equilibrium is attained, the upward and downward forces are balanced and the ball moves at a constant velocity. That is, the falling ball reaches a terminal velocity (v_t) when the acceleration due to the force of gravity is exactly compensated by the friction of the fluid on the ball.

$$\frac{dv}{dt} = 0 \tag{2.41}$$

In the Stoke's region, the drag coefficient is:

$$c_D = \frac{24}{Re} \tag{2.42}$$

Substituting Eqs. (2.41) and (2.42) into Eq. (2.40) gives:

$$\frac{\pi D_p^3 \rho_p g}{6} = \frac{\pi D_p^3 \rho_f g}{6} + \frac{6\pi D_p \mu v_t}{2} \tag{2.43}$$

$$\Rightarrow \mu = \frac{D_p^2 (\rho_p - \rho_f) g}{18 v_t} \tag{2.44}$$

If the terminal velocity of the ball is calculated, it is possible to determine the dynamic viscosity of the fluid. Falling ball viscometers are more suitable for viscous fluids where the terminal velocity is low. Stoke's law applies when the diameter of the ball is so much smaller than the diameter of the tube through which it is falling. Thus, there is no effect of the wall on the rate of fall of the ball. If the tube diameter is 10 times the ball diameter, wall effects can be neglected.

The larger the ball, the faster it falls. Therefore, it is necessary to select the diameter of ball small enough to fall at a rate that can be measured with some degree of accuracy. This method is not suitable for opaque fluids.

The rising bubble viscometer operates on the same principle as the falling ball viscometer. In this case, a bubble of air is allowed to rise through a column of liquid. Rising time for a certain distance is correlated to viscosity.

Example 2.4. To determine the viscosity of sunflower oil, a falling ball viscometer was used. Viscometer has a tube length of 10 cm and its ball has a diameter of 0.68 mm. Oil and the ball have densities of 921 kg/m^3 and 2420 kg/m^3, respectively. If it takes 44.5 s for the ball to fall from the top of the tube, calculate the viscosity of the oil.

Solution:

Terminal velocity is:

$$v_t = \frac{L}{t}$$

$$= \frac{0.1 \, m}{44.5 \, s}$$

$$= 0.0022 \, m/s$$

Then, viscosity can be calculated using Eq. (2.44):

$$\Rightarrow \mu = \frac{D_p^2 (\rho_p - \rho_f) g}{18 v_t} \tag{2.44}$$

$$\mu = \frac{(0.68 \times 10^{-3} \, m)^2 (2420 \, kg/m^3 - 921 \, kg/m^3)(9.81 \, m/s^2)}{(18)(0.0022 \, m/s)} = 0.172 \, Pa \cdot s$$

2.3.4   Rotational Viscometers

In rotational viscometers, the sample is sheared between the two parts of the measuring device by means of rotation. In agitation, the shear rate is proportional to the rotational speed. It is possible to measure the shear stress as the shear rate is changed. In addition, a sample can be sheared for as long as desired. Therefore, rotational viscometers are the best for characterization of non-Newtonian and time-dependent behavior. There are different forms of these viscometers, as described in the following sections.

2.3.4.1   Concentric Cylinder (Coaxial Rotational) Viscometers

This type of viscometer consists of two annular cylinders with a narrow gap between them (Fig. 2.10). The fluid to be measured is placed in the gap. Either the inner (Searle system) or the outer (Couette system) cylinder is rotated. Although the Searle system is more common, the word Couette is mostly used to refer to any kind of concentric cylinder system. For both Couette and Searle systems, the equations relating rotation to torsion (M) are the same. By changing the shear rate or shear stress, it is possible to obtain viscosity measurements over a range of shearing conditions on the same sample. It can be used for both Newtonian and non-Newtonian foods.

The following assumptions should be made in developing the mathematical relationships (Steffe, 1996):

1. Flow is laminar and steady.
2. Radial and axial velocity components are zero.
3. The test fluid is incompressible.
4. The temperature is constant.
5. End effects are negligible.
6. There is no slip at the wall of the instrument.

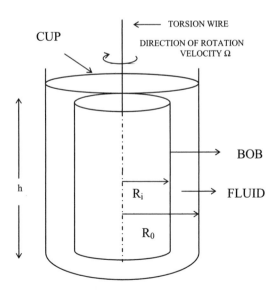

Figure 2.10   Concentric cylinder viscometer.

Consider a test fluid placed in Searle type viscometer. When the inner cylinder rotates at a constant angular velocity (Ω) and the outer cylinder is stationary, the instrument measures the torque (M) required to maintain this constant angular velocity of the inner cylinder. The opposing torque comes from the shear stress exerted on the inner cylinder by the fluid. Making a force balance:

$$M = 2\pi r h r \tau = 2\pi h r^2 \tau \tag{2.45}$$

where r is any location in the fluid and h is the height of the cylinder. Solving Eq. (2.45) for the shear stress:

$$\tau = \frac{M}{2\pi h r^2} \tag{2.46}$$

This equation shows that the shear stress is not constant over the gap between the two concentric cylinders but decreases in moving from the inner cylinder of radius, R_i, to the outer cylinder of radius, R_0. The shear stress at the inner cylinder can be written as:

$$\tau_i = \frac{M}{2\pi h R_i^2} \tag{2.47}$$

To determine the shear rate, the linear velocity (v) is expressed in terms of angular velocity (ω) at r:

$$\dot{\gamma} = -\frac{dv}{dr} = -r\frac{d\omega}{dr} \tag{2.48}$$

The shear rate is a function of shear stress, and the functional relationship between shear stress and shear rate is shown by:

$$\dot{\gamma} = -r\frac{d\omega}{dr} = f(\tau) \tag{2.49}$$

An expression for the differential of the angular velocity yields:

$$d\omega = -\frac{dr}{r}f(\tau) \tag{2.50}$$

An expression for r can be determined from Eq. (2.45):

$$r = \left(\frac{M}{2\pi h \tau}\right)^{1/2} \tag{2.51}$$

Equation (2.51) is differentiated with respect to τ and substituted into Eq. (2.50) after substituting the value of torque from Eq. (2.45):

$$d\omega = \frac{1}{2}f(\tau)\frac{d\tau}{\tau} \tag{2.52}$$

Integrating Eq. (2.52) over the fluid in the annulus gives the general expression for the angular velocity of the inner cylinder as a function of the shear stress in the gap:

$$\int_{\Omega}^{0} d\omega = \frac{1}{2}\int_{\tau_i}^{\tau_0} f(\tau)\frac{d\tau}{\tau} \tag{2.53}$$

However, the solution of Eq. (2.53) depends on $f(\tau)$, which depends on the behavior of the fluid.

If the fluid is Newtonian:

$$\dot{\gamma} = f(\tau) = \frac{\tau}{\mu} \tag{2.54}$$

Substituting Eq. (2.54) into Eq. (2.53) gives:

$$\Omega = -\frac{1}{2\mu} \int_{\tau_i}^{\tau_0} d\tau = \frac{1}{2\mu}(\tau_i - \tau_0) \tag{2.55}$$

Substituting Eq. (2.46) into Eq. (2.55) and rearrangement gives Margules equation:

$$\Omega = \frac{M}{4\pi\mu h} \left(\frac{1}{R_i^2} - \frac{1}{R_o^2} \right) \tag{2.56}$$

For a power law fluid:

$$\dot{\gamma} = f(\tau) = \left(\frac{\tau}{k} \right)^{1/n} \tag{2.57}$$

Substituting Eq. (2.57) into Eq. (2.53) gives:

$$\Omega = -\frac{1}{2} \int_{\tau_i}^{\tau_0} \left(\frac{\tau}{k} \right)^{1/n} \frac{d\tau}{\tau} = \frac{n}{2k^{1/n}} \left[(\tau_i)^{1/n} - (\tau_0)^{1/n} \right] \tag{2.58}$$

Substituting Eq. (2.46) into Eq. (2.58), an expression for the power law fluid is obtained:

$$\Omega = \frac{n}{2k^{1/n}} \left(\frac{M}{2\pi h R_i^2} \right)^{1/n} \left[1 - \left(\frac{R_i}{R_o} \right)^{2/n} \right] \tag{2.59}$$

When studying liquid foods, the simple shear, Newtonian, or power law approximations are often used (Steffe, 1996).

(a) Simple Shear Approximation. If there is a very small gap between the cylinders compared to the radius, shear stress can be taken as constant. Assuming a uniform shear rate across the gap:

$$\dot{\gamma}_i = \frac{\Omega R_i}{R_0 - R_i} = \frac{\Omega}{\alpha - 1} \tag{2.60}$$

where $\alpha = {R_o}/{R_i}$

When calculating shear rates with Eq. (2.60), a corresponding average shear stress should be used:

$$\tau_{\text{ave}} = \frac{1}{2}(\tau_i + \tau_o) = \frac{M(1 + \alpha^2)}{4\pi h R_o^2} \tag{2.61}$$

(b) Newtonian Approximation. Shear stress at the inner cylinder can be calculated using Eq. (2.47). An equation for the shear rate at the inner cylinder can be derived by substituting Eq. (2.45) into the Margules Eq. (2.56) and expressing shear stress in terms of shear rate using Newton's law of viscosity:

$$\dot{\gamma}_i = 2\Omega \left(\frac{\alpha^2}{\alpha^2 - 1} \right) \tag{2.62}$$

(c) Power Law Approximation. Shear stress at the inner cylinder can be calculated using Eq. (2.47). An equation for the shear rate at the inner cylinder can be derived by substituting Eq. (2.45) into Eq. (2.59) and expressing shear stress in terms of shear rate using power law equation:

$$\dot{\gamma}_i = \left(\frac{2\Omega}{n}\right)\left(\frac{\alpha^{2/n}}{\alpha^{2/n} - 1}\right) \tag{2.63}$$

where the flow behavior index n is:

$$n = \frac{d(\ln \tau_i)}{d(\ln \Omega)} = \frac{d(\ln M)}{d(\ln \Omega)} \tag{2.64}$$

2.3.4.2 Cone and Plate Viscometers

The operating principle for cone and plate viscometers is similar to that for concentric cylinder viscometers. The system consists of a circular plate and a cone with radius R with its axis perpendicular to the plate and its vertex in the plane of the surface of the plate (Fig. 2.11). Usually, the cone is rotated at a known angular velocity (Ω). The fluid is placed in the gap between the cone and plate and transmits torque to the plate. If the angle θ between the cone and plate is small ($<5°$), shear stress and shear rate are uniform over the fluid (Steffe, 1996). This type of viscometer is suitable for shear thinning fluids and plastic fluids. It is well suited for testing small samples. End effects do not play a role. However, at higher shear rates, there is a danger of low-viscosity fluids being thrown out of the gap. Keeping the temperature constant during measurement is more difficult for cone and plate geometries than for concentric cylinders. Materials containing large particles cannot be studied because of the small gap between the cone and the plate.

To develop a mathematical relationship, the shear rate at r may be written as:

$$\dot{\gamma} = \frac{r\Omega}{r \tan \theta} = \frac{\Omega}{\tan \theta} \tag{2.65}$$

This indicates that the shear rate is constant throughout the gap. If the angle is small, $\tan \theta$ can be approximated to be equal to θ.

Since the shear rate is constant in the gap, the shear stress is also constant. To develop an expression for shear stress, the differential torque on an annular ring of thickness (dr) is considered and integrated over the radius:

$$\int_0^M dM = \int_0^R (2\pi r \, dr) \, \tau \, r \tag{2.66}$$

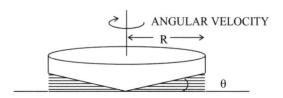

Figure 2.11 Cone and plate viscometer.

Table E.2.5.1 Torque Values as a Function of Angular Velocity for Semisolid Vanilla Dairy Dessert in Cone and Plate Viscometer

Angular Velocity (rad/min)	Torque (N · m)
1.04	4.66×10^{-4}
15.7	7.62×10^{-4}
31.4	9.60×10^{-4}
52.2	11.73×10^{-4}
73.1	13.56×10^{-4}
104.4	15.96×10^{-4}

Then shear stress is:

$$\tau = \frac{3M}{2\pi R^3} \tag{2.67}$$

Shear stress and shear rate can be calculated for different angular velocities, cone angles, and cone radius from Eqs. (2.65) and (2.67) and using torque data. Then rheological properties can be determined after selection of a specific model:

For a Newtonian fluid:

$$\frac{3M}{2\pi R^3} = \mu \left(\frac{\Omega}{\theta} \right) \tag{2.68}$$

For a power law fluid:

$$\frac{3M}{2\pi R^3} = k \left(\frac{\Omega}{\theta} \right)^n \tag{2.69}$$

Example 2.5. A semisolid vanilla dairy dessert was examined for its rheological properties at 25°C by using a cone and plate viscometer with 50-mm diameter and 1° angle. The dessert was sheared in the viscometer at increasing angular velocity and the measured torque values as a function of angular velocity are given in Table E.2.5.1. Examine Newtonian, power law, and Herschel-Bulkley models to find the expression that best describes the flow behavior of the dessert.

Solution:

Using angular velocity (Ω) data, the shear rate ($\dot{\gamma}_w$) is calculated from Eq. (2.65):

$$\dot{\gamma}_w = \frac{\Omega}{60 \tan \theta}$$

(1/60) is the conversion factor to convert angular velocity from rad/min to rad/s. Using torque (M) data, shear stress τ_w is calculated from Eq. (2.67):

$$\tau_w = \frac{3M}{2\pi R^3} \tag{2.67}$$

For a Newtonian fluid, τ_w versus ($\dot{\gamma}_w$) is plotted (Fig. E.2.5.1).

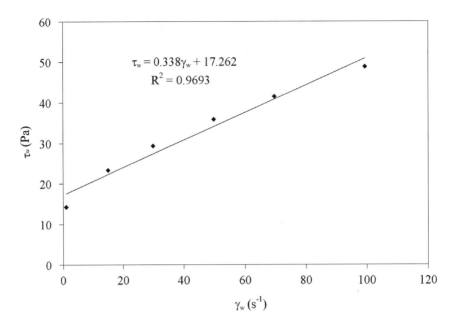

Figure E.2.5.1   Shear stress versus shear rate plot for dairy dessert.

The fluid is not Newtonian since the intercept of the graph shows that the fluid has a yield stress.

For a power law fluid, a logarithmic plot of shear stress versus shear rate ($\ln \tau_w$ versus $\ln \dot{\gamma}_w$) is drawn (Fig. E.2.5.2). This plot yields a straight line with a slope of n and intercept of $\ln k$.

$$\ln \tau_w = 2.581 + 0.260 \ln \dot{\gamma}_w \ (r^2 = 0.954)$$

The slope of the model gives n, which is 0.260.

From the intercept which is $\ln k = 2.581$, the value of k is calculated as 13.210 Pa $\cdot$ (s)^n

Thus, the power law expression is:

$$\tau_w = 13.210 \, \dot{\gamma}_w^{0.260}$$

The Herschel-Bulkley expression is given in Eq. (2.17):

$$\tau_w = \tau_0 + k \, (\dot{\gamma}_w)^n \tag{2.17}$$

To determine the value for τ_0, shear stress at the wall (τ_w) is plotted with respect to shear rate ($\dot{\gamma}_w$), and τ_0 is found to be approximately 13.5 Pa by extrapolation.

To find the model constants, the Herschel-Bulkley expression is linearized as:

$$\ln(\tau_w - \tau_0) = \ln k + n \ln \dot{\gamma}_w$$

The plot of $\ln(\tau_w - \tau_0)$ versus $\ln \dot{\gamma}_w$ is shown in Fig. E.2.5.3.

When linear regression is used, the equation of $\ln(\tau_w - \tau_0) = 0.264 + 0.733 \ln \dot{\gamma}$ is obtained with a coefficient of determination (r^2) of 0.998. From the intercept which is $\ln k = 0.264$, the value of k is 1.302 Pa(s)^n. Thus, the Herschel-Bulkley expression is:

$$\tau_w = 13.5 + 1.302 \, \dot{\gamma}_w^{0.733}$$

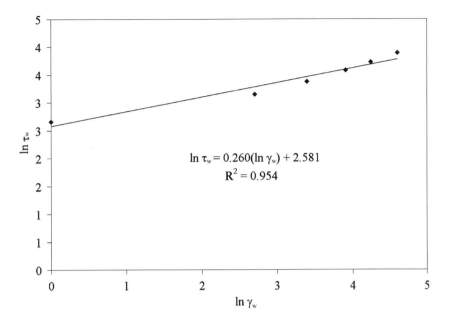

Figure E.2.5.2 Power law model for a dairy dessert.

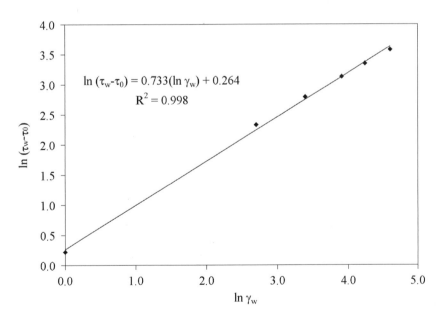

Figure E.2.5.3 Herschel-Bulkley model for a dairy dessert.

Since the highest r^2 value was obtained for the Herschel-Bulkley model, it can be concluded that it describes the flow behavior of the dessert best.

2.3.4.3 Parallel Plate Viscometers

The system consists of two parallel plates separated at a distance h from each other. The sample is placed in the gap between two parallel plates in this type of viscometer. One of the plates is rotated at a known angular velocity (Ω) while the other one is stationary.

In contrast to cone and plate systems, the shear rate is not constant in the fluid during deformation but changes as a function of distance from the center r in the parallel plate apparatus as shown in the following equation:

$$\dot{\gamma} = \Omega \frac{r}{h} \tag{2.70}$$

The angular velocity (Ω) is in 1/s. Equation (2.66) can also be written for parallel plate systems but the integration is more complicated since shear stress changes as a function of r. The integration gives:

$$\tau_R = \frac{M}{2\pi R^3} \left[3 + \frac{d \ln M}{d \ln \dot{\gamma}_R} \right] \tag{2.71}$$

which is similar in form to the Rabinowitsh-Mooney equation (Eq. 2.35).

For a Newtonian fluid:

$$\tau = \mu \dot{\gamma} = \mu \frac{\Omega r}{h} \tag{2.72}$$

Substituting this into Eq. (2.66) and integrating gives:

$$\frac{2M}{\pi R^3} = \mu \left(\frac{\Omega R}{h} \right) \tag{2.73}$$

Similarly for power law fluids:

$$\frac{M(3+n)}{2\pi R^3} = k \left(\frac{\Omega R}{h} \right)^n \tag{2.74}$$

and

$$\tau_R = \frac{M(3+n)}{2\pi R^3} \tag{2.75}$$

Example 2.6. A food company is trying to find a new soup formulation in their Research & Development department. The rheological properties of the soup formulation are determined for this purpose. A parallel plate viscometer with $R = 25$ mm and $h = 0.7$ mm is used in rheological studies. The torque versus angular velocity data given in Table E.2.6.1 are taken at 15°C for this formulation. Determine the power law constants describing the flow behavior of the soup.

Solution:

Using angular velocity (Ω) data, shear rate ($\dot{\gamma}_R$) can be calculated from Eq. (2.70) as:

$$\dot{\gamma} = \frac{(\Omega/60)R}{h}$$

Table E.2.6.1 Torque Values as a Function of Angular Velocity for a Soup Formulation in Parallel Plate Viscometer

Torque (N · m)	Angular Velocity (rad/min)
0.000821	2.3
0.000972	4
0.001190	7
0.001723	18
0.002977	52

(1/60) is the conversion factor to convert angular velocity from radian/min to radian/s. Using angular velocity data shear rate values were calculated. For non-Newtonian fluids shear stress can be calculated from torque by Eq. (2.71):

$$\tau_R = \frac{M}{2\pi R^3} \left(3 + \frac{d \ln M}{d \ln \dot{\gamma}_R}\right) \tag{2.71}$$

When linear regression was applied to the data of $(\ln M)$ versus $(\ln \dot{\gamma}_R)$, the slope of the equation is found as 0.412. Inserting this value into Eq. (2.71) in place of $\dfrac{d \ln M}{d \ln \dot{\gamma}_R}$, the expression for τ_R is determined:

$$\tau_R = \frac{M}{2\pi R^3} (3 + 0.412)$$

Using torque data, shear stress values were calculated. Then, $(\ln \tau_R)$ versus $(\ln \dot{\gamma}_R)$ was plotted to determine the power law constants (Fig. E.2.6.1).

From the linear regression, a straight line was obtained (Fig. E.2.6.1) and the following model was found:

$$\ln \tau_R = 0.412 \ln \dot{\gamma}_R + 3.173$$

From the intercept, which is $\ln k = 3.173$, the value of k is calculated as 23.879 Pa · $(s)^n$

Thus, the power law model for soup is:

$$\tau_R = 23.879 \, \dot{\gamma}_R^{0.412}$$

2.3.4.4 Single-Spindle Viscometers (Brookfield Viscometer)

A spindle attached to the instrument with a vertical shaft is rotated in the fluid and the torque necessary to overcome the viscous resistance is measured. Different spindles are available in various sizes which may be rotated at different speeds. A suitable spindle and a rotational speed for a particular fluid are selected by trial and error. This device gives the viscosity of Newtonian fluids directly since it is calibrated with Newtonian oils. The steady-state deflection is noted and a conversion chart is provided to estimate the apparent viscosity under the test conditions. It is possible to determine apparent viscosity at different speeds (shear rates), but since it is not possible to state the exact shear

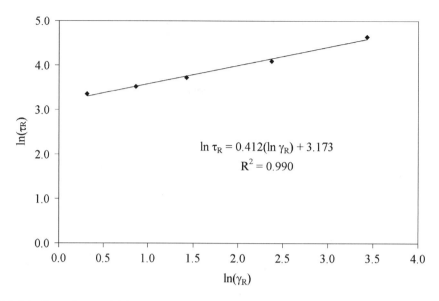

Figure E.2.6.1   Power law model for a soup.

rate that the fluid is subjected to, shear stress–shear rate data cannot be presented. The results are normally presented in the form of apparent viscosity against rotational speed. A range of calibration fluids is available for checking the accuracy of the equipment. It is possible to determine whether the fluid is time dependent or not by using this viscometer.

2.3.5   Other Types of Viscometers

2.3.5.1   Vibrational (Oscillation) Viscometer

Vibrational viscometers use the principle of surface loading whereby the surface of an immersed probe generates a shear wave that dissipates in the surrounding medium. The power required to maintain constant amplitude of oscillation is proportional to the viscosity of the fluid. The container should be large enough so that the shear forces do not reach the wall and reflect back to the probe. Measurements depend on the ability of the surrounding fluid to damp the probe vibration. The damping characteristic of a Newtonian fluid is a function of the fluid viscosity and the density. These types of viscometers are popular as in-line instruments for process control systems.

2.3.5.2   Bostwick Consistometer

The instrument is a simple device usually made of stainless steel and consists of two compartments separated by a spring loaded gate. The first compartment has dimensions of 5 × 5 × 3.8 cm. The second compartment is 5 cm wide, 24 cm long, and approximately 2.5 cm high. Its floor is graduated in 0.5 cm increments. The gate is lowered and the first compartment is filled by the sample. The test begins by pressing the trigger that releases the gate and fluid flows under the influence of gravity into the second compartment consisting of an inclined trough. The movement of fluid down the trough reflects the fluid properties. The distance of travel from the starting gate is reported in cm after a specified time (5–30 s). If fluid motion produces a curved surface, the travel distance of the leading edge is considered. It is used as a quality control tool in food industry for tomato paste, applesauce, etc.

The following relation was obtained between Bostwick readings and apparent viscosity of fluids (McCarthy & Seymour, 1994):

$$\text{Bostwick reading} = \left(\frac{\eta}{\rho}\right)^{-0.2} \tag{2.76}$$

2.4 DEFORMATION OF MATERIAL

It is important to discuss stress and strain thoroughly to understand rheology of foods. Stress is defined as force per unit area. It is generally expressed in Pa (N/m^2). Stress can be categorized into two groups: normal stress and shear stress. The difference between these two stresses depends on the area that the force acts.

Normal stress (σ) is defined as the force applied perpendicular to the plane per unit area. Pressure is an example of normal stress. Normal stress can be tensile or compressive depending on whether it tends to stretch or to compress the material on which it acts (Fig. 2.12a). In shear stress, the stress acts tangential to the surface. Shear stress (τ) is defined as the force applied parallel to the plane per unit area (Fig. 2.12b).

Strain is the unit change in size or shape of a material referred to its original size or shape when a force is applied.

$$\text{Strain} = \frac{\text{Extension}}{\text{Original Length}} \tag{2.77}$$

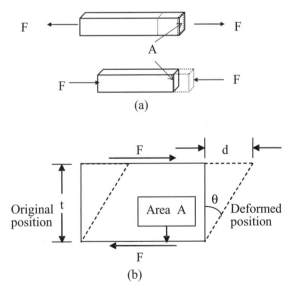

(a)

(b)

Figure 2.12 (a) Tensile and compressive normal stress and (b) shear stress.

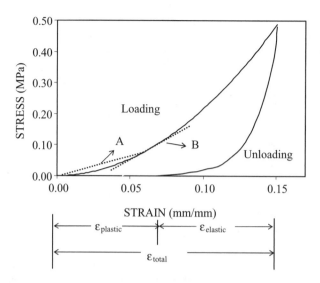

Figure 2.13   Stress–strain curve for compression of a food material.

Strain, like stress, can be categorized into two groups: normal strain and shear strain. Normal strain (ε) is the change in length per unit length in the direction of the applied normal stress:

$$\varepsilon = \frac{\Delta L}{L} \tag{2.78}$$

Shear strain (γ) is defined as the change in the angle formed between two planes that are orthogonal prior to deformation as a result of the application of stress (Fig. 2.12b):

$$\gamma = \tan\theta = {}^d\!/_t \tag{2.79}$$

Stresses and strains can also be described as either dilatational or deviatoric. A dilatational stress or strain causes change in volume while a deviatoric stress or strain is the one that results in change in shape. The normal stresses and strains that usually cause volume changes are called dilatational while "pure" shear stresses, which tend to distort the shape of a sample while causing negligible changes in volume, are called deviatoric. Dilatation can be calculated from the initial (V_0) and final (V_f) volumes of the sample:

$$\text{Dilatation} = \frac{V_f - V_0}{V_0} \tag{2.80}$$

A stress–strain curve for compression of a food sample is similar to that shown in Fig. 2.13. Change in strain as a function of stress during loading and unloading can be seen in the figure. Strain that is not recovered during unloading is called plastic strain, while strain that is recovered is called elastic strain. The ratio of plastic strain to total strain when a material is loaded to a certain load and then unloaded is called the degree of plasticity. Similarly, the ratio of elastic strain to total strain is defined as degree of elasticity. When a stress is applied to a purely elastic solid, it will deform finitely but then it will return to its original position after the stress is removed. Material showing elastic behavior is known as a Hookean solid.

Foods that follow Hookean's law are dry pasta, egg shells, and hard candies when subjected to small strains (e.g., < 0.01) (Steffe, 1996). Large strains produce brittle fracture or nonlinear behavior. For a Newtonian fluid, all the energy input necessary to make it flow at a given rate is dissipated as heat while for a Hookean solid the energy necessary to deform is stored as a potential energy that is fully recoverable.

The strain energy density at a given strain is the area under the loading curve in the stress–strain curve. The area under the unloading curve is called **resilience**. It is the energy per unit volume recovered as the force is removed from the sample. The greater the resilience, the more energy will be recovered. The difference between the strain energy density and the resilience is called the **hysteresis**.

The ratio of stress to strain is known as **modulus** while the ratio of strain to stress is known as **compliance**. Different types of moduli are defined for a Hookean solid.

Young's modulus or Modulus of elasticity (E) is defined as the ratio of normal stress (σ) to normal tensile or compressive strain (ε).

$$E = \frac{\sigma}{\varepsilon} \tag{2.81}$$

In the case of food materials, the apparent modulus of elasticity is used to relate stress to strain since stress–strain curve is not linear and no single E value is obtained. It can be defined using secant or tangent definition (Fig. 2.13). In the secant definition, the apparent modulus of elasticity is the ratio of the stress to strain at a given point (A). In the tangent definition, it is defined as the slope of the stress–strain curve at a given point on the curve (B). It can be calculated using the central difference approximation of the first derivative at a point.

Shear modulus or modulus of rigidity (G) is used to describe the relationship between the shear stress and shear strain.

$$G = \frac{\tau}{\gamma} \tag{2.82}$$

If the force is applied from all directions which results in a volume change, modulus is called **bulk modulus (K)**.

$$K = \frac{\text{Average normal stress}}{\text{Dilatation}} = \frac{\text{Pressure change}}{\text{Volume change / Original volume}} \tag{2.83}$$

When a sample is subjected to uniaxial compression in one direction, it may expand in the other directions. **Poisson's ratio (μ)** is defined as the ratio of the strain in the direction perpendicular to the applied force to the strain in the direction of the applied force.

$$\mu = \frac{\text{Change in width per unit width}}{\text{Change in length per unit length}} = \frac{\Delta D/D}{\Delta L/L} \tag{2.84}$$

Bioyield point is defined as the point at which an increase in deformation is observed with a decrease or no change of force. In some agricultural products, the presence of this bioyield point is an indication of initial cell rupture.

Rupture point is a point on the stress–strain or force-deformation curve at which the axially loaded specimen ruptures under a load. Rupture point corresponds to a failure in the macrostructure of the specimen while bioyield point corresponds to a failure in the microstructure of the sample.

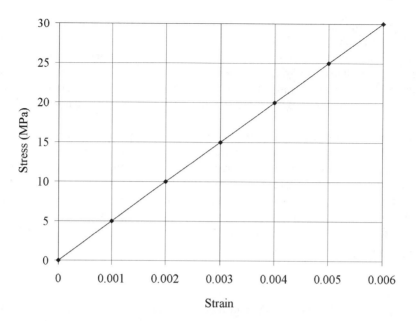

Figure E.2.7.1   Stress versus strain data obtained from a tensile test applied on semolina fibers.

Example 2.7. Dry commercial semolina fibers with a diameter of 1.65 mm are used to examine the rheological properties of dry spaghetti.

(a) A tensile test is applied on fibers of 150 mm length and the result is given in Fig. E.2.7.1. Determine the value of the modulus of elasticity.

(b) What is the Poisson ratio if the fibers exhibit a diameter change of 2.43×10^{-3} mm under the stress of 15 MPa?

Solution:

(a) Modulus of elasticity is defined as:

$$E = \frac{\sigma}{\varepsilon} \tag{2.81}$$

From the slope of the curve, E is found to be 5000 MPa.

(b) Poisson ratio is $\mu = \dfrac{\Delta D/D}{\Delta L/L}$ \hfill (2.84)

From the graph at 15 MPa stress, the strain, $\varepsilon = \dfrac{\Delta L}{L}$, is read as 0.003.

Inserting $\dfrac{\Delta L}{L}$ into Eq. (2.84), Poisson ratio (μ) can be calculated as:

$$\mu = \frac{2.43 \times 10^{-3}/1.65}{0.003} = 0.490$$

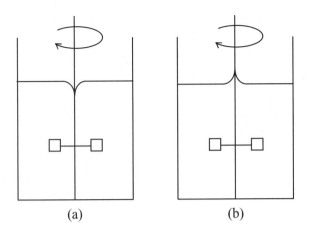

(a)   (b)

Figure 2.14   (a) Vortex formation observed in viscous fluids. (b) Weissenberg effect observed in viscoelastic fluids.

2.5   VISCOELASTIC BEHAVIOR

When a force is applied to a viscous fluid, it will start to deform and this deformation is proportional with the magnitude of force applied. It deforms continuously until the force is removed so that it cannot return to its original position. Viscous fluids generally exhibit viscosity while solids exhibit elasticity. Some foods show both viscous and elastic properties which are known as viscoelastic materials. The typical food example for viscoelastic fluids is wheat flour dough. Dairy cream, ice cream mix, marshmallow cream, cheese, and most gelled products are also viscoelastic foods. There is no simple constant for viscoelastic materials such as modulus because the modulus will change with respect to time.

When a viscous fluid is agitated, the circular motion causes a vortex. If a viscoelastic fluid is stirred by a rotating rod it tends to climb the rod, which is known as the **Weissenberg effect** (Fig. 2.14). You might have observed this effect while mixing a cake batter or bread dough at home. This is due to the production of a normal force acting at right angles to the rotational forces, which in turn acts in a horizontal plane. The rotation tends to straighten out the polymer molecules in the direction of rotation but the molecules attempt to return to their original position.

When a Newtonian fluid emerges from a long, round tube into the air, the emerging jet will normally contract. It may expand to a diameter of 10% to 15% larger than the tube diameter at low Reynolds numbers. Normal stress differences present in a viscoelastic fluid, however, may cause jet expansion (called die swell) which are two or more times the diameter of the tube. In addition, highly elastic fluids may exhibit a tubeless siphon effect (Steffe, 1996).

Another phenomenon observed in the viscoelastic material is called the **recoil phenomenon.** When the flow of viscoelastic material is stopped, tensile forces in the fluid cause particles to move back. However, viscous fluids stay where they are when their motion is stopped (Steffe, 1996). This phenomenon is illustrated in Fig. 2.15.

There are three different methods to study viscoelastic materials: stress relaxation test, creep test, and dynamic test.

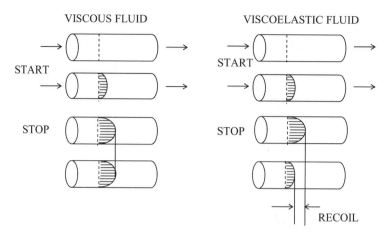

Figure 2.15 Recoil phenomenon in viscous and viscoelastic fluids. [From Steffe, J.F. (1996). *Rheological Methods in Food Process Engineering,* 2nd ed. East Lansing, MI: Freeman Press (available at www.egr.msu.edu/~steffe/freebook/offer.html).]

2.5.1 Stress Relaxation Test

If food materials are deformed to a fixed strain and the strain is held constant, the stress required to maintain this strain decreases with time. This is called stress relaxation. In this test, stress is measured as a function of time as the material is subjected to a constant strain. This test can be conducted in shear, uniaxial tension, or uniaxial compression. Figure 2.16 shows stress relaxation curves for elastic, viscous, and viscoelastic materials.

As can be seen in Fig. 2.16, ideal viscous substances relax instantaneously but no relaxation is observed in ideal elastic materials. Viscoelastic materials relax gradually and stop depending on the

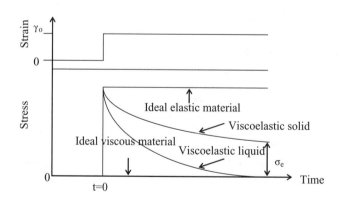

Figure 2.16 Stress relaxation curves for elastic, viscous and viscoelastic materials [From Steffe, J.F. (1996). *Rheological Methods in Food Process Engineering*, 2nd ed. East Lansing, MI: Freeman Press (available at www.egr.msu.edu/~steffe/freebook/offer.html).]

molecular structure of the material. Stress in viscoelastic solids will decay to an equilibrium stress σ_e, which is greater than zero but the residual stress in viscoelastic liquids is zero.

The stresses in liquids can relax more quickly than those in solids because of the higher mobility of liquid molecules. Relaxation time is very short for liquids, which is 10^{-13} s for water while it is very long for elastic solids. For viscoelastic material, relaxation time is 10^{-1}–10^6 s (van Vliet, 1999).

In recent years, the stress relaxation test has been performed to study viscoelastic behavior of sago starch, wheat flour, and sago–wheat flour mixture (Zaidul, Karim, Manan, Azlan, Norulaini, & Omar, 2003), potato tuber (Blahovec, 2003), cooked potatoes (Kaur, Singh, Sodhi, & Gujral, 2002), wheat dough (Li, Dobraszczyk, & Schofield, 2003; Safari-Ardi & Phan-Thien, 1998), and osmotically dehydrated apples and bananas (Krokida, Karathanos, & Maroulis, 2000).

Studies on weak and strong wheat flour dough showed that stress relaxation tests at high strain values could differentiate dough from high-protein and low-protein wheat cultivars (Safari-Ardi & Phan-Thien, 1998). Stress relaxation measurements on wheat flour dough and gluten in shear showed that relaxation behavior of dough could be explained by two relaxation processes: a rapid relaxation over 0.1 to 10 s and a slower relaxation occurring over 10 to 10,000 s (Bohlin & Carlson, 1980). The rapid relaxation process is related to small polymers that relax rapidly and longer relaxation time is associated with high molecular weight polymers found within gluten. Similarly, stress relaxation behavior of wheat dough, gluten, and gluten protein fraction obtained from biscuit flour showed two relaxation processes: a major peak at short times and a second peak at times longer than 10 s (Li et al., 2003). Many researchers showed that a slower relaxation time is associated with good baking quality (Bloksma, 1990; Wang & Sun, 2002).

2.5.2 Creep Test

If a constant load is applied to biological materials and if stresses are relatively large, the material will continue to deform with time. This is known as creep. In a creep test, an instantaneous constant stress is applied to the material and the resulting strain is measured as a function of time. There is a possibility of some recovery of the material when the stress is released as the material tries to return to its original shape.

Creep test can be performed in uniaxial tension or compression. The creep curves for elastic, viscous, and viscoelastic materials are shown in Fig. 2.17. For liquids, strain increases with time in a steady manner and the observed stress will be constant with time. Ideal viscous material shows no recovery since it is affected linearly with stress. Viscoelastic material such as bread dough shows partial recovery. They show a nonlinear response to strain since they have the ability to recover some structure by storing energy.

Results of a creep test are expressed as creep compliance ($J = \gamma/\tau$). For ideally elastic materials, creep compliance is constant while it changes as a function of time for viscoelastic materials.

Recently, viscoelastic properties of wheat dough having different strengths were determined using a creep test (Edwards, Peressini, Dexter, & Mulvaney, 2001). A creep time of 10,000 s was enough to reach the steady-state flow for all of the dough with different strengths. When wheat flour dough was analyzed by a creep-recovery test, a maximum recovery strain of wheat dough was correlated to some of the parameters provided by using a mixograph, farinograph, and texture analyzer (Wang & Sun, 2002). When a creep recovery test was applied to biscuit dough, there was an increase in percentage of recovery as aging time was increased (Pederson, Kaack, Bergsøe, & Adler-Nissen, 2004). This shows that dough is becoming less extensible but more recoverable as aging increases. Maximum strain and recovery were strongly affected by wheat cultivar. The rheological measurements from creep test and

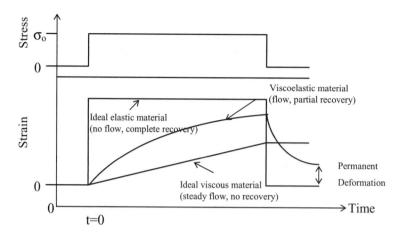

Figure 2.17   Creep and recovery curves for elastic, viscous and viscoelastic materials. [From Steffe, J.F. (1996). *Rheological Methods in Food Process Engineering*, 2nd ed. East Lansing, MI: Freeman Press (available at www.egr.msu.edu/~steffe/freebook/offer.html).]

oscillation tests showed that rice dough with 1.5% and 3.0% HPMC had similar rheological properties to that of wheat flour dough (Sivaramakrishnan, Senge, & Chattopadhyay, 2004).

2.5.3   Dynamic Test (Oscillatory Test)

In dynamic tests, either rate is controlled (stress is measured at a constant strain) or stress is controlled (deformation is measured at a constant stress amplitude). That is, materials are subjected to deformation or stress which varies harmonically with time. Usually, a sinusoidal strain is applied to the sample, causing some level of stress to be transmitted through the material. Then, the transmitted shear stress in the sample is measured (Fig. 2.18).

Concentric cylinder, cone, and plate or parallel viscometers are suitable for this purpose. This test is suitable for undisturbed viscoelastic materials as a function of time. Both elastic and viscous components can be obtained over a wide range of time. The main disadvantage of this test is that it can be used in the region in which stress is proportional with strain. Otherwise interpretation of data is hard. Moreover, the breakdown of structure of material may occur during the experiment (van Vliet, 1999).

The magnitude and the time lag (phase shift) of the transmission depend on the viscoelastic nature of the material. Much of the stress is transmitted in highly elastic materials while it is dissipated in frictional losses in highly viscous ones. The time lag is large for highly viscous materials but small for highly elastic materials.

A storage modulus (G') that is high for elastic materials and loss modulus (G'') that is high for viscous materials are defined as follows:

$$G' = \frac{\tau_0 \cos\theta}{\gamma_0} \tag{2.85}$$

$$G'' = \frac{\tau_0 \sin\theta}{\gamma_0} \tag{2.86}$$

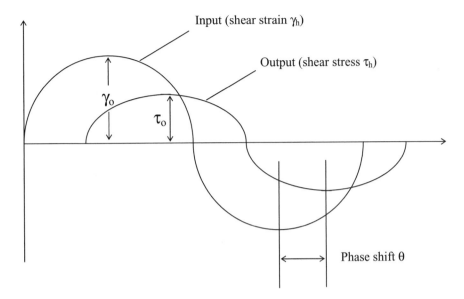

Figure 2.18 Harmonic shear stress versus strain for a viscoelastic material in dynamic test.

2.6 EXTENSIONAL FLOW

Many food processes involve extensional deformation. Pure extensional flow does not involve shearing and sometimes called shear free flow. In extensional flow, molecular orientation is in the direction of the flow field since there are no competing forces to cause rotation. Extensional flow causes maximum stretching of molecules, producing a chain tension that may result in a large resistance to deformation.

Dough processing is an important food process in which extensional flow is significant. Another example for extensional flow is extrusion, which involves a combination of shear and extensional flow. Formation of carbon dioxide gas during fermentation of bread dough involves extensional deformation. The extensograph is an important instrument used to study dough rheology which gives resistance to extension and stretchability.

There are three types of extensional flow: uniaxial, planar, and biaxial (Fig. 2.19). During uniaxial extension, material is stretched in one direction with a size reduction in the other two directions. In planar extension, a material is stretched in the x_1 direction with a corresponding decrease in x_2 while width in the x_3 direction remains unchanged. In biaxial extension the flow produces a radial tensile stress.

There are many test methods that measure the uniaxial extensional properties of dough such as the Brabender Extensograph and the Stable Microsystems Kieffer dough and gluten extensibility rig (Dobraszczyk & Morgenstern, 2003). However, these methods do not give rheological data in units of stress and strain since sample geometry is not defined or measured, and dimensions change nonuniformly during testing. Therefore, it is impossible to determine stress, strain, modulus, or viscosity by these methods.

The commonly used methods to measure biaxial properties of foods are inflation methods and compression between flat plates using lubricated surfaces (Chatraei, Macosko, & Winter, 1981; Dobraszcyk & Vincent, 1999). The alveograph operates under the principle of biaxial extension.

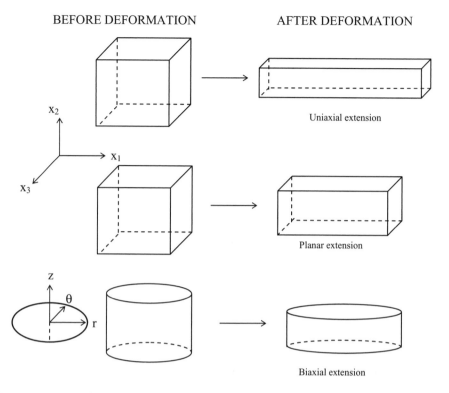

BEFORE DEFORMATION   AFTER DEFORMATION

Uniaxial extension

Planar extension

Biaxial extension

Figure 2.19   Uniaxial, planar and biaxial extension. [From Steffe, J.F. (1996). *Rheological Methods in Food Process Engineering*, 2nd ed. East Lansing, MI: Freeman Press (available at www.egr.msu.edu/~steffe/freebook/offer.html).]

2.7   MECHANICAL MODELS

To understand the rheological behavior of viscoelastic foods, some mechanical models are used. These models consist of springs and dashpots.

2.7.1   Elastic (Spring) Model

A Hookean spring is shown in Fig. 2.20. The deformation distance of spring (x) varies linearly with the force acting on the system (F):

$$F = k\,x \tag{2.87}$$

where k is the spring constant. The spring is considered as an ideal solid element obeying Hooke's

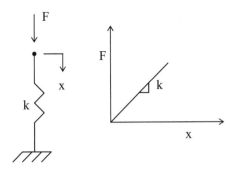

Figure 2.20   Elastic (spring) model.

law:

$$\tau = G\gamma \tag{2.88}$$

2.7.2   Viscous (Dashpot) Model

A dashpot is shown in Fig. 2.21. The rate of extension varies linearly with the force acting on the system:

$$F = C\frac{dx}{dt} \tag{2.89}$$

where C is a constant proportional to the diameter of the holes.

The dashpot is considered as an ideal fluid element obeying Newton's law in which force is proportional to rate of extension:

$$\tau = \mu\frac{d\gamma}{dt} \tag{2.90}$$

$$\tau = \mu\dot{\gamma} \tag{2.91}$$

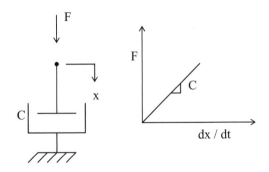

Figure 2.21   Viscous (dashpot) model.

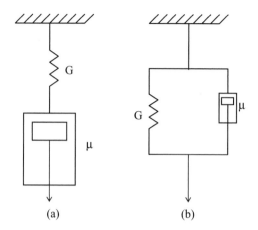

Figure 2.22   (a) Maxwell model and (b) Kelvin model.

2.7.3   Combination Models

Springs and dashpots can be connected in different ways to express the behavior of viscoelastic materials. The most common models are the Maxwell and Kelvin-Voigt models (Fig. 2.22). In Maxwell model the spring and dashpot are connected in series whereas they are connected in parallel in the Kelvin model.

2.7.3.1   Maxwell Model

The Maxwell model has been used to interpret stress relaxation of viscoelastic liquids, especially polymeric liquid. Since the arrangement is a series arrangement in Maxwell model (Fig. 2.22a), the total shear strain can be expressed as the summation of strain in the spring and dashpot:

$$\gamma = \gamma_{spring} + \gamma_{dashpot} \tag{2.92}$$

Differentiating Eq. (2.92) with respect to time and using Eqs. (2.88) and (2.91), the following equation was obtained:

$$\frac{d\gamma}{dt} = \dot{\gamma} = \frac{1}{G}\left(\frac{d\tau}{dt}\right) + \frac{\tau}{\mu} \tag{2.93}$$

This equation can be expressed as:

$$\mu\dot{\gamma} = \tau + \lambda_{rel}\left(\frac{d\tau}{dt}\right) \tag{2.94}$$

where λ_{rel} is the relaxation time and defined as:

$$\lambda_{rel} = \frac{\mu}{G} \tag{2.95}$$

If a constant strain γ is used, shear rate, $\dot{\gamma}$ becomes zero and after integrating Eq. (2.94) with respect

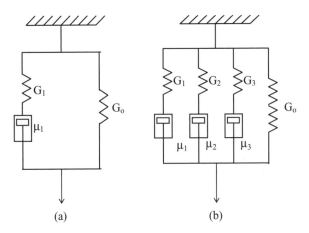

Figure 2.23 Maxwell elements in parallel with spring (a) single Maxwell element and a free spring and (b) three Maxwell element and a free spring.

to time, Maxwell model can be expressed for constant (G/μ) as:

$$\tau = \tau_0 \exp\left(-\frac{t}{\lambda_{\text{rel}}}\right) \tag{2.96}$$

Equation (2.96) describes the gradual relaxation of stress (from τ_0 to zero) after the application of a sudden strain. This relationship helps determining the relaxation time (λ_{rel}), which is the time it takes for the stress to decay to $1/e$ (36.8%) of its initial value.

The Maxwell model is not suitable for many viscoelastic materials since it does not include an equilibrium stress (τ_e). It can be modified by connecting a parallel spring to the Maxwell element (Fig. 2.23a). The stress relaxation equation described by this three-element model is:

$$\tau(t) = \tau_e + (\tau_0 - \tau_e)\exp\left(-\frac{t}{\lambda_{rel}}\right) \tag{2.97}$$

where

$$\tau_e = G_0\gamma_0 \tag{2.98}$$

The relaxation time is defined in the standard Maxwell portion of the model as:

$$\lambda_{\text{rel}} = \frac{\mu_1}{G_1} \tag{2.99}$$

The number of Maxwell elements can be increased to fit the experimental stress relaxation data. In Fig. 2.23b, three Maxwell elements in parallel with a free spring can be seen. In the study of Kaur et al. (2002), a seven-element Maxwell model (three Maxwell elements and a free spring) was in good agreement with experimental data for cooked potatoes. The stress relaxation data of osmotically dehydrated apples and bananas were modeled using two terms Maxwell model (Krokida et al., 2000). Osmotic pretreatment decreased relaxation time of dehydrated samples, showing that sugar gain increases viscous nature but decreases elastic nature of fruits.

In viscoelastic fluids, stress is a function of the strain similar to a solid and it is also a function of strain rate like a liquid. The degree that a fluid can return to its original position depends on its elastic and viscous characteristics. A dimensionless group, the Deborah number, is important for viscoelastic materials. For any fluid Deborah number (De) is:

$$De = \frac{\lambda_{rel}}{\theta} \tag{2.100}$$

where λ_{rel} is the characteristic relaxation time of the material and θ is the characteristic time of the deformation process.

The Deborah number represents the ratio of duration of fluid memory to the duration of the deformation process and is used as a measure of degree of viscoelasticity. If De < 1, that is, if the relaxation time of the liquid is less than the characteristic deformation time, the fluid appears to be more viscous than elastic. On the contrary, if De > 1, it appears to be more elastic than viscous.

2.7.3.2 Kelvin-Voigt Model

Creep behavior can be described by the Kelvin-Voigt model. This model contains a spring and a dashpot connected in parallel (Fig. 2.22b). Therefore, it is possible to express all strains as equal to each other.

$$\gamma = \gamma_{spring} = \gamma_{dashpot} \tag{2.101}$$

Total shear stress caused by the deformation is sum of the individual stresses:

$$\tau = \tau_{spring} + \tau_{dashpot} \tag{2.102}$$

Equations (2.88) and (2.91) can be substituted into Eq. (2.102). Then:

$$\tau = G\gamma + \mu\dot{\gamma} \tag{2.103}$$

The variation of stress with time will be zero when the material is allowed to flow after being subjected to a constant shear stress (τ_0), in creep. Substituting $\tau = \tau_0$, into Eq. (2.103) and integrating as:

$$\int_0^t dt = \mu \int_0^\gamma \frac{d\gamma}{\tau_0 - G\gamma} \tag{2.104}$$

gives

$$\gamma = \frac{\tau_0}{G} \left(1 - \exp\left(-\frac{t}{\lambda_{ret}}\right) \right) \tag{2.105}$$

where λ_{ret} is the retardation time, which is unique for any substance and is expressed as:

$$\lambda_{ret} = \frac{\mu}{G} \tag{2.106}$$

This equation shows that strain changes from zero to the maximum value (τ_0/G) asymptotically. Retardation time (λ_{ret}) is the time for the delayed strain to reach approximately 63.2% ($1 - 1/e$) of the final value.

Time to achieve the maximum strain is delayed in viscoelastic material while retardation time is zero and the maximum strain is obtained immediately with application of stress for a Hookean solid.

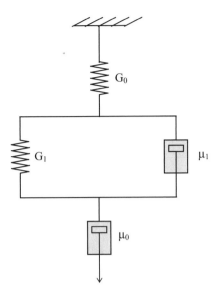

Figure 2.24 Burger model.

2.7.3.3 Burger Model

The Kelvin model shows excellent retardation but this is not the case in many foods. Therefore, the Burger model shown in Fig. 2.24 is proposed, which is the series combination of the Kelvin and Maxwell models:

$$\gamma = \frac{\tau_0}{G_0} + \frac{\tau_0}{G_1}\left(1 - \exp\left(\frac{-t}{\lambda_{\text{ret}}}\right)\right) + \frac{\tau_0 t}{\mu_0} \tag{2.107}$$

where $\lambda_{\text{ret}} = \dfrac{\mu_1}{G_1}$ is the retardation time for the Kelvin part of the model, and μ_0 is the Newtonian viscosity of the free dashpot.

This model shows an initial elastic response due to the free spring, retarded elastic behavior related to the parallel spring-dashpot combination, and Newtonian type of flow after long periods of time due to a free dashpot.

The Burger model can also be expressed in terms of creep compliance by dividing Eq. (2.107) by the constant stress (τ_0):

$$\frac{\gamma}{\tau_0} = \frac{1}{G_0} + \frac{1}{G_1}\left(1 - \exp\left(\frac{-t}{\lambda_{\text{ret}}}\right)\right) + \frac{t}{\mu_0} \tag{2.108}$$

Writing the equation as a creep compliance function:

$$J = J_0 + J_1\left(1 - \exp\left(\frac{-t}{\lambda_{\text{ret}}}\right)\right) + \frac{t}{\mu_0} \tag{2.109}$$

where J_0 is the instantaneous compliance and J_1 is the retarded compliance. The sum of J_0 and J_1 is called the steady-state compliance. The compliance and recovery (recoil) curves showing compliance parameters for the Burger model can be seen in Fig. 2.25.

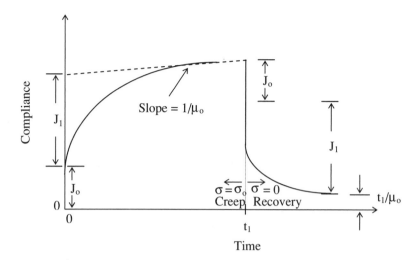

Figure 2.25 The compliance and recovery (recoil) curves showing compliance parameters for Burger model. [From Steffe, J.F. (1996). *Rheological Methods in Food Process Engineering*, 2nd ed. East Lansing, MI: Freeman Press (available at www.egr.msu.edu/~steffe/freebook/offer.html).

Viscoelastic properties of different wheat dough having different strengths were analyzed by creep compliance test and the data showed a good agreement with Burger model (Edwards et al., 2001).

Example 2.8. Gellan gel is examined for its stress relaxation behavior (Table E.2.8.1). Assuming that equilibrium is reached within 10 min and data fit the three-element Maxwell model, determine the relaxation parameters for a gellan gel.

Table E.2.8.1 Stress Relaxation Data for Gellan Gel

Time (min)	Stress (kPa)
0.00	45.0
0.10	39.0
0.25	35.0
0.50	29.0
1.00	22.5
1.50	18.5
2.00	16.0
3.00	13.5
4.00	12.0
5.00	10.8
6.00	10.5
7.00	10.2
8.00	10.0
10.00	9.5

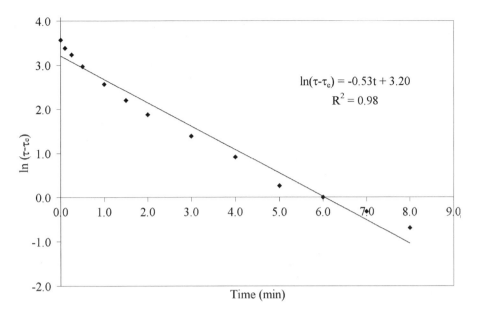

Figure E.2.8.1 Stress relaxation curve for gellan gel.

Solution:

The three-element Maxwell model is:

$$\tau(t) = \tau_e + (\tau_0 - \tau_e)\exp\left(-\frac{t}{\lambda_{rel}}\right) \tag{2.97}$$

Modifying the equation and taking the natural logarithm of both sides:

$$\ln(\tau(t) - \tau_e) = \ln(\tau_0 - \tau_e) - \left(\frac{t}{\lambda_{rel}}\right)$$

From the given data, $\tau_e = 9.5$ kPa (Table E.2.8.1).

$\ln(\tau - \tau_e)$ versus time is plotted as shown in Fig. E.2.8.1.

From the linear regression, the slope, which is equal to $(-1/\lambda_{rel})$, is determined to be -0.53. Thus, λ_{rel} is 1.887 min. From the intercept, which is $\ln(\tau_0 - \tau_e)$, $(\tau_0 - \tau_e)$ is found to be 24.62 kPa. Then, the three-element Maxwell equation for stress relaxation of a gellan gel is:

$$\tau(t) = 9.5 + (24.62)\exp\left(-\frac{t}{1.887}\right)$$

Example 2.9. The creep analysis results were obtained for wheat flour dough by applying a constant stress of 50 Pa for 60 s on the sample. The data are given in Table E.2.9.1. Using creep compliance data, determine the viscoelastic parameters, G_0, G_1, μ_1, and μ_0 of a Burger model.

Solution:

Burger model is:

$$\gamma = \frac{\tau_0}{G_0} + \frac{\tau_0}{G_1}\left(1 - \exp\left(\frac{-t}{\lambda_{ret}}\right)\right) + \frac{\tau_0 t}{\mu_0} \tag{2.107}$$

Table E.2.9.1
Deformation of Wheat
Flour Dough

Time (s)	Deformation
0	0.0060
5	0.0095
10	0.0140
15	0.0160
20	0.0180
25	0.0188
30	0.0195
35	0.0203
40	0.0210
45	0.0218
50	0.0225
55	0.0233
60	0.0240

The Burger model in terms of creep compliance is:

$$J(t) = J_0 + J_1 \left(1 - \exp\left(\frac{-t}{\lambda_{\text{ret}}} \right) \right) + \frac{t}{\mu_0} \tag{2.109}$$

First, the deformation data are converted into compliance data using the constant stress of 50 Pa. Then, creep compliance versus time graph is plotted to determine the model parameters (Fig. E.2.9.1). The instantaneous compliance, J_0 is determined from the raw data as 0.00012 Pa^{-1}. Then:

$$J_0 = \frac{1}{G_0}$$

$$\Rightarrow G_0 = 8333.33 \, \text{Pa}$$

Then, using the straight line portion of the $(J - J_0)$ versus time curve (the last eight data points) (Fig. E.2.9.2), linear regression analysis yields retardation compliance, J_1 and μ_0, from the intercept and slope, respectively:

$$(J - J_0) = J_1 + \frac{t}{\mu_0}$$

Note that J_1 reflects the fully equilibrated Kelvin element, making the exponential term of the Eq. (2.109) equal to zero.

The slope of the straight line portion of the graph is $\dfrac{1}{\mu_0} = 2.99 \times 10^{-6}$

$$\Rightarrow \mu_0 = 334395 \, \text{Pa} \cdot \text{s}$$

and from the intercept of the straight portion of the graph, J_1 is found to be 0.00018 Pa^{-1}. G_1 can be calculated from J_1 as 5555.6 Pa.

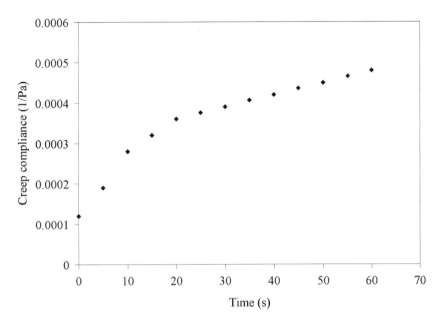

Figure E.2.9.1   Creep compliance curve for wheat flour.

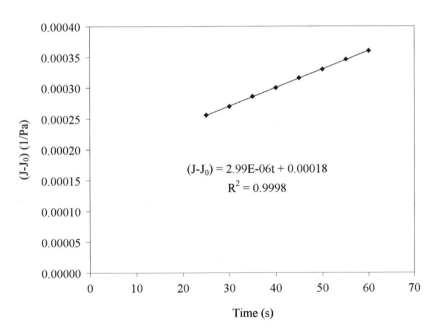

Figure E.2.9.2   Linear regression of the straight line portion of creep compliance data.

Using the exponential portion of the data (initial period), the retardation time is determined from the linear regression analysis over $J < J_1 + J_0$:

$$\ln\left(1 - \frac{J - J_0}{J_1}\right) = -\frac{t}{\lambda_{ret}}$$

$$\lambda_{ret} = \frac{\mu_1}{G_1} = 5.21\,s$$

Then, $\mu_1 = 28944.68$ Pa·s. Substituting the model parameters into Eq. (2.109), the Burger model may be expressed as:

$$J(t) = 0.00012 + 5555.6\left(1 - \exp\left(\frac{-t}{5.21}\right)\right) + \frac{t}{334395}$$

2.8   TEXTURE OF FOODS

Texture is one of the most important quality characteristics of foods. Foods have different textural properties. These differences are caused by inherent differences due to the variety difference, differences due to maturity, and differences caused by processing methods.

Food texture can be evaluated by sensory or instrumental methods. Sensory methods need a taste panel containing trained panelists. It is hard to repeat the results. Instrumental methods are less expensive and less time consuming as compared to sensory methods. There are various instrumental methods to determine the texture of foods. More detailed discussion about texture can be found in the chapter written by Dobraszczyk and Vincent (1999).

2.8.1   Compression

Compression (deformation) test measures the distance that a food is compressed under a standard compression force or the force required to compress a food a standard distance. This test can be compared to the squeezing of bread by the consumers to be sure that bread is fresh. The sensory description of this test is softness or firmness.

According to the AACC method, bread firmness can be determined by using compression principle (AACC, 1988). In this test, a Universal Testing machine fitted with 36 mm diameter of aluminum plunger with cross-head speed of 100 mm/min and a chart speed of 50 mm/min is used. One slice of bread having a thickness of 25 mm or two breads each having a thickness of 12.5 mm are used. The force to achieve 25% compression in Newtons is read off the chart. This test can also be used for other baked products. Figure 2.26 shows the variation of firmness values of high ratio cakes baked in microwave and conventional ovens during storage determined by AACC method. Firm texture is one of the problems in microwave baked products.

2.8.2   Snapping-Bending

This test measures the force required to bend or snap brittle foods such as biscuits or crackers. The sample is laid across two vertical rails that support it in horizontal position. A third bar mounted above the sample and equidistant between the supporting rails is lowered until the sample breaks and the

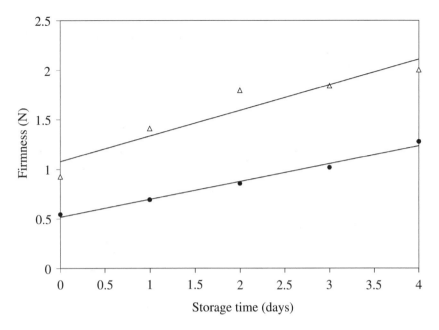

Figure 2.26 Variation of firmness of cakes baked in conventional and microwave oven during storage. (●), Conventional oven; (△), microwave oven. [From Seyhun, N. (2002). *Retardation of Staling of Microwave Baked Cakes*. MS thesis, Middle East Technical University: Ankara, Turkey.

force is measured. The force required to snap the sample depends on the strength and the dimensions of the sample (Bourne, 1990).

The three-point bending test is the most commonly used snapping-bending test. This test is used for biscuit and chocolate bars that are homogeneous and long compared to their thickness and width. Samples should have a length to thickness ratio of at least 10 for this test. The upper part of test piece is compressed and the lower part is elongated during bending. In between there is a neutral axis (Fig. 2.27).

Three-point bending has loading at a single central point for a beam supported at either end. The deflection at the center of the beam produced by a given force is expressed as:

$$\rho = \frac{FL^3}{48EI} \qquad (2.110)$$

where ρ is the deflection at the center of the beam, F is the force, L is the length between supports, E is Young's modulus, and I is the second moment of area.

For a rectangular section beam having length of w and thickness of t, $I = \dfrac{wt^3}{12}$; for a solid square beam $I = \dfrac{t^4}{12}$, for a solid cylindrical beam having a radius of r, $I = \dfrac{\pi r^4}{4}$ (van Vliet, 1999).

Substituting the expression of I for a rectangular section beam into Eq. (2.110):

$$\rho = \frac{FL^3}{4Ewt^3} \qquad (2.111)$$

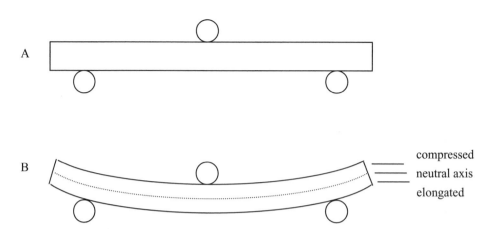

Figure 2.27   Illustration of three-point bending test.

The snapping force can be expressed as (Bruns & Bourne, 1975):

$$F = \frac{2}{3}\sigma_{max}\frac{wt^2}{L}$$ (2.112)

where σ_{max} is the maximum stress.

For bars with cylindrical cross sections such as pretzels and bread sticks, the snapping equation becomes (Bourne, 1990):

$$F = \sigma_{max}\frac{\pi r^3}{L}$$ (2.113)

2.8.3   Cutting Shear

The Pea Tenderometer, which was introduced in 1937, works via the principle of cutting shear. It consists of a grid of blades rotated at constant speed through a second grid of blades. As the peas are cut by the blades, the maximum force is measured. This instrument is still used to determine the maturity of peas at harvest.

The Kramer Shear press was also developed to determine the texture of peas. It is widely used to determine the texture of fruits and vegetables. A typical system contains 10 shear blades that are 3.2 mm thick and separated by a distance equal to thickness. The sample holder is filled with the food. Shear blades are forced through the material until they pass through the bars in the bottom of the sample container. Force on the ram holding the blades is measured over time and correlated to the firmness of the product.

2.8.4   Puncture

Puncture test measures the force required to push the probe into the food and expressed as firmness or hardness of the product. It is used mostly for fruits, gels, vegetables, and some dairy or meat products.

The puncture test is not widely used on cereal products since hard baked products are susceptible to fracture when subjected to this test.

Puncture force is proportional to both area and perimeter of the probe and compression and shear properties of food. The relationship is shown by Bourne (1966) as:

$$F = K_c A + K_s P + C \tag{2.114}$$

where F is the puncture force, K_c is the compression coefficient of food, K_s is the shear coefficient of food, A is the probe area, P is the probe perimeter, and C is a constant.

For most of the foods, the constant C is zero within the limits of experimental error. For some foods such as hard baked foods the value of K_s is close to zero, which simplifies Eq. (2.114) to:

$$F = K_c A \tag{2.115}$$

2.8.5 Penetration

Penetrometers were originally designed to measure the distance that a cone or a needle sinks into a food such as margarine or mayonnaise under the force of gravity for a standard time. This is a simple and relatively inexpensive apparatus used for determination of spreadability of butter (Walstra, 1980). The penetration depth depends on weight, cone angle, falling height, and properties of test materials. The cone will first deform the material and at large deformation the material may yield or fracture. This flow contains both shear and elongational flow components (Fig. 2.28). Penetration depth is determined by a combination of elastic modulus in shear and compression, yield or fracture stress, and shear and elongational viscosity (van Vliet, 1999).

2.8.6 Texture Profile Analysis

Texture Profile Analysis (TPA) compresses a bite-sized piece of food (usually 1 cm cube) twice to simulate the chewing action of the teeth. Compression is usually 80% of the original length of the

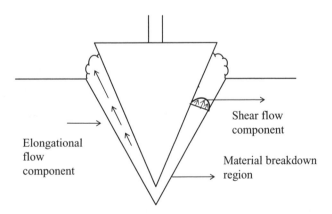

Figure 2.28   Illustration of penetrometer. [From Van Vliet, T. Rheological classification of foods and instrumental techniques for their study. In A.J. Rosenthal (Ed.), *Food Texture Measurement and Perception* (pp. 65–98). New York: Aspen. Copyright © (1999) with permission from Springer.]

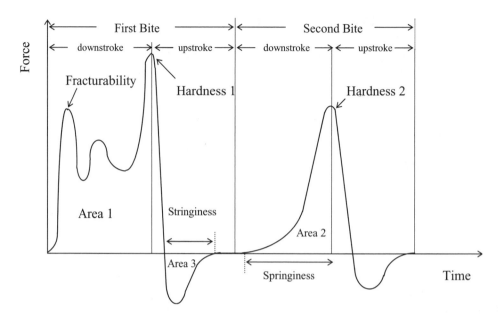

Figure 2.29   Generalized texture profile.

sample. As a result of TPA, sensory properties such as gumminess, cohesiveness, and so forth can be determined objectively. Texture analyzers are used to obtain texture profile analysis.

The force curve generated as a function of time is known as a texture profile. Since the instrument compresses the sample twice, two positive and two negative curves are obtained (Fig. 2.29). Peak forces and areas under the curves are used to determine various properties of foods like fracturability, hardness, cohesiveness, adhesiveness, springiness, gumminess, and chewiness.

Fracturability (brittleness) is defined as the force at the first significant break in the first positive bite area.

Hardness is defined as the peak force during the first compression cycle.

Cohesiveness is defined as the ratio of the second positive bite area to the first positive bite area.

Adhesiveness is defined as the negative force area for the first bite representing the work required to pull the plunger away from the food.

Springiness (elasticity) is defined as the height to which the food recovers during the time that elapses between the end of the first bite and start of the second bite (distance or length of the compression cycle during the second bite).

Gumminess is the product of hardness and cohesiveness. In sensory terms, it is the energy required to disintegrate a semisolid food so that it is ready for swallowing.

Chewiness is the product of gumminess and springiness. In sensory terms, it is known as the energy required for chewing a solid food until it is ready for swallowing.

Texture profiles for chicken nuggets and bread crumbs can be seen in Fig. 2.30. Chicken nuggets show very high fracturability and hardness values, indicating the crispness of the product. On the other hand, bread crumbs have a softer texture.

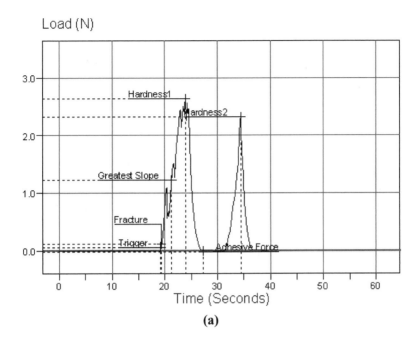

(a)

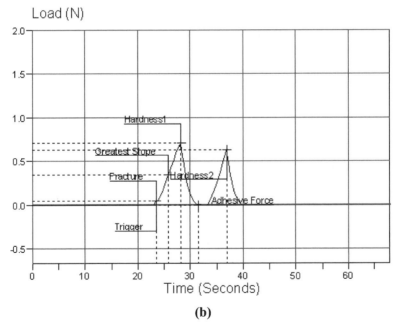

(b)

Figure 2.30 Texture profiles for different food examples: (**a**) chicken nugget; (**b**) bread.

2.9   DOUGH TESTING INSTRUMENTS

Knowledge of dough rheology will give information about the final product quality. The instruments to investigate dough behavior are the farinograph, mixograph, extensograph, and alveograph. These instruments measure power input during dough development caused by either mixing action or extensional deformation.

2.9.1   Farinograph and Mixograph

Both the farinograph and the mixograph are torque measuring devices that provide empirical information about mixing properties of flour by recording resistance of dough to mixing. These instruments differ in their mixing action. In the farinograph there is a kneading type of mixing. There are two Z-shaped blades that rotate at different speeds at different directions. The mixograph involves a planetary rotation of vertical pins (lowered into the dough) attached to the mixing bowl.

The farinograph gives information about mixing properties of flour by recording the resistance of the dough to mixing blades during prolonged mixing. The output of this instrument is known as a farinogram, which is a consistency versus time curve. Consistency is expressed as Brabender units (BU). The shape of the farinogram is interpreted in terms of dough development time, tolerance to overmixing, stability, and optimum water absorption (Fig. 2.31).

The time required for dough to reach the maximum consistency is called peak time. It is also called dough development time. Tolerance of the dough to overmixing can be expressed as stability, mixing tolerance index (MTI) and departure time. Stability is the duration at which the dough consistency is ≥500 BU. It is an indication of flour strength. Higher values indicate stronger dough. Mixing tolerance index is the change in dough consistency 5 min after it reaches its maximum value. The time for the consistency to decrease below 500 BU is known as departure time. Water absorption (%) is obtained by measuring the amount of water required to produce dough having a consistency of 500 BU.

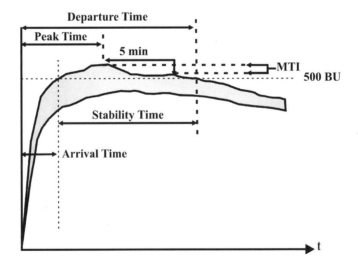

Figure 2.31   A general farinogram.

Table 2.1 Farinograph Analysis of Dough Containing Hydrocolloids[a]

	Water Absorption (%)	DDT (min)	Stability (min)	MTI (BU)
Control	54.9	2.8	10.5	40
Alginate	59.5	9.0	15.5	30
K-carragenan	56.5	2.5	5.8	60
Xanthan	56.7	11.5	20.0	20
HPMC	57.9	2.8	9.8	40

[a]From Rosell, C.M., Rojas, J.A, & de Barber, C.B. Influence of hydrocolloids on dough rheology and bread quality. *Food Hydrocolloid, 15 (1)*, 75–81. Copyright © (2001), with permission from Elsevier.

The shape of the farinogram curve depends on the variety of wheat, environmental conditions, and type of the flour produced during milling. Hard wheat or bread type flours have longer stability values than flours from soft wheat or cake type flour.

Hydrocolloids are usually added to bread formulations to improve dough handling properties, to increase quality of fresh bread, and to extend shelf-life of bread. Table 2.1 shows the effect of hydrocolloid addition on farinograph measurements (Rosell, Rojas, & de Barber, 2001). Water absorption increased when hydrocolloids was added. The extent of increase was dependent on the structure of hydrocolloid added. Dough development time (DDT) was affected differently by each hydrocolloid. The strongest dough was obtained by addition of alginate or xanthan which, was reflected in high stability and low MTI values.

In the mixograph, torque is recorded while a fixed amount of flour and water is mixed. Output of this instrument is known as mixogram. The mixograph is more complicated than the farinograph since there is no predetermined optimum consistency that will be taken as a reference. Therefore, other methods should be used to determine the amount of water required to produce dough of optimum absorption. For a mixograph, the tolerance of flour can be seen in weakening angle (W), the area under the curve, the height of the curve at a specified time after peak, and the angle between the ascending and descending portions of the curve which is known as tolerance angle (T) (Fig. 2.32).

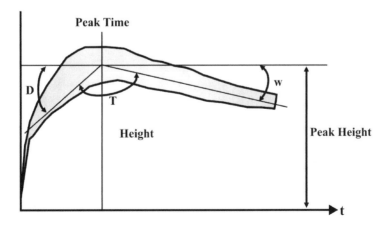

Figure 2.32 A general mixogram.

Tolerance angle (T) is obtained by drawing a line from center of the curve at its peak down the center of the curve in both directions. A large tolerance angle indicates more tolerant flour. Weakening angle (W) is formed by drawing a line for center of the curve at its peak down the descending portion of the curve and a line horizontal to the baseline through the center of the curve at its maximum height. The size of the angle (W) is inversely related to mixing tolerance. The development angle (D) is formed by a line drawn horizontal to the baseline through the center of the curve at its maximum height and a line drawn through the center of the ascending part of the curve.

2.9.2   Extensograph and Alveograph

Measurement of the rheological properties of the dough after mixing is made using an extensograph and an alveograph, which measure stress–strain relationships, thus providing information about elasticity.

An extensograph gives information about dough resistance to stretching and extensibility (Fig. 2.33). It measures the force to pull a hook through a rod-shaped piece of dough. The extensograph gives resistance to constant deformation after 50 mm of stretching ($R50$) and extensibility. The resistance to stretching (B) is related to elastic properties and the extensibility (C) is related to the viscous component. The ratio of resistance to stretching to extensibility (B/C) is a good indicator of the balance between elastic and viscous components of the dough. The area under the extensogram (A) indicates energy and is related to the absolute levels of elastic and viscous components of the dough. A combination of good resistance and good extensibility results in desirable dough properties. It is also possible to determine stretching and extensibility values with a texture analyzer if Kieffer dough and a gluten extensibility rig is used.

An alveograph, which is also called the Chopin Entensograph, provides similar information by measuring the pressure required to blow a bubble in a sheeted piece of dough. An alveogram is shown in Fig. 2.34. The advantage of an alveograph over an extensograph is the mode of expansion. There is a constant rate of extension in only one direction with an extensograph but an alveograph expands the

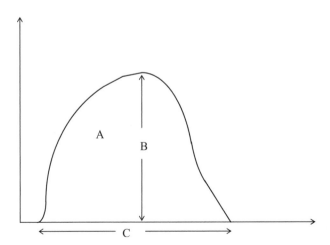

Figure 2.33   Example of an extensogram.

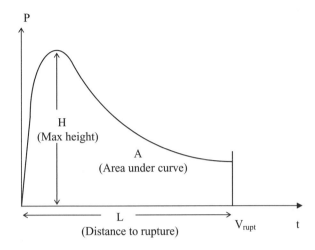

Figure 2.34 Example of an alveogram.

dough in two directions and the rate of expansion varies as the bubble grows. This action resembles the action on a piece of dough during fermentation and early stages of baking.

The equations used in an alveograph are:

$$P = (1.1)H \tag{2.116}$$

where P is overpressure (mm)

$$G = \sqrt{V_{\text{rupt}}} \tag{2.117}$$

where G is swelling index (ml)

$$W = \frac{1.32V}{L\,A} \tag{2.118}$$

where

W = deformation energy (10^{-4} J),
V = volume of air (mL),
L = abscissa at rupture (mm).

P is an indicator of resistance to deformation or resistance to extension. The elastic component is in the maximum height of the curve and length of the curve is related to viscous portion. The P/L ratio is the balance of elastic to viscous components of the dough. Length is highly correlated with loaf volume. Dough resistance to deformation is a predictor of the ability of the dough to retain gas. Alveograph data have been used to calculate biaxial extensional viscosity.

The disadvantage of both instruments is the relative complexity of testing procedures. Since both methods require that dough should be formed into a reproducible shape, skilled operators are required (Spies, 1990).

2.9.3 Amylograph

An amylograph is a torsion viscometer that records changes in viscosity of starch as temperature is raised at constant rate of approximately 1.5°C/min. Viscosity of starch suspensions is generally recorded in BU units as the temperature rises from 30 to 95°C. An amylograph is generally used to determine gelatinization characteristics of various starches. In Europe, an amylograph is widely used to predict baking performance of rye flours and to detect excessive amounts of flour from spouted grains (Pomeranz, 1987). In the United States, it is used to control malt supplementation.

PROBLEMS

2.1. Intrinsic viscosity is a characteristic property of a polymer in a specific solvent that measures the hydrodynamic volume occupied by a polymer molecule. The gelling ability of whey proteins provides important textural and water holding properties in many foods. In a study, whey protein concentrates with different concentrations were studied for their intrinsic viscosity values. Intrinsic viscosity was measured using Cannon-Fenske capillary viscometers immersed in a constant temperature bath at 25°C. The recorded time values for protein concentrates to pass from the upper mark to the lower mark of viscometer are given in Table P.2.1.1. The efflux time for the solvent, which is water, was measured to be 370 s.

Hint: Since the protein solutions are very dilute, their densities can be assumed the same as the density of water.

(a) Calculate the relative, specific, and reduced viscosity values for each protein concentrate at different concentrations.

(b) Determine k_1(Huggins constant) and intrinsic viscosity values for protein.

2.2. A rising bubble viscometer consists of a glass vessel that is 30 cm deep. It is filled with a liquid at constant temperature having a density of 1260 kg/m^3. The time necessary for a bubble having a diameter of 1 cm and a density of 1.2 kg/m^3 to rise 20 cm up the center of column of liquid is measured as 4.5 s. Calculate the viscosity of the liquid.

2.3. Coconut milk is extracted from fresh coconut flesh. The rheological behavior of coconut milk containing 30% fat has been studied and has been found to obey the power law. Table P.2.3.1 shows the consistency coefficient (k) and flow behavior index (n) values obtained at different temperatures (data taken from Simuang, Chiewchan, & Tansakul, 2004).

Table P.2.1.1 Timing Results for Different Protein Concentrations

Protein Concentration (g/mL)	Time (s)
0.0005	380.0
0.0010	389.5
0.0015	398.5
0.0020	405.3
0.0025	413.5

Table P.2.3.1 Consistency Coefficient and Flow Behavior Index of Coconut Milk for Different Temperatures

Temperature (°C)	k (Pa.s^n)	n
25	5.05×10^{-2}	0.851
70	4.38×10^{-2}	0.739
80	3.80×10^{-2}	0.756
90	3.49×10^{-2}	0.758

Table P.2.4.1 Torque Versus Angular Velocity Data for Cake Batter

Angular Velocity (rad/min)	Torque (N · m)
0	0
3600	0.00043
7200	0.00059
10,800	0.00075
14,400	0.00088
18,000	0.00093
21,600	0.00097
25,200	0.00101
28,800	0.00104
32,400	0.00107
36,000	0.00109

Calculate the Arrhenius parameters for the effect of temperature on the apparent viscosity of the coconut milk sample at a shear rate of 300 s^{-1}.

2.4. Table P.2.4.1 shows torque versus angular velocity data obtained at 25°C for cake batter using cone and plate viscometer with 0.05 m diameter and 3° angle. Determine the flow behavior and the model of the cake batter.

2.5. Table P.2.5.1 shows torque versus angular velocity data taken for a sauce at constant temperature using a parallel plate viscometer with two plates having a radius of 25 mm and separated by a gap of 0.7 mm. Determine the flow behavior of the sauce.

2.6. A student taking a FDE 490 Food Engineering Research course, has collected flow rate versus pressure drop data for a 1.5% aqueous solution of sodium carboxymethylcellulose (CMC) at room temperature using a 45-cm-long capillary tube viscometer with 2-mm diameter during his studies. The night before the FDE 490 final, while preparing his presentation, the student spilled his coffee over his data sheet and it became impossible to read many parts of the sheet, except the flow rate data. Fortunately, he had already found that the data fitted the power law model representing the fluid behavior as:

$$\tau_w = 8.14 \, (\dot{\gamma}_w)^{0.414} \qquad (r^2 = 0.999)$$

Table P.2.5.1 Torque Versus
Angular Velocity Data for a Sauce

Torque (N · m)	Angular Velocity (rad/min)
0.000821	2.3
0.000972	4
0.001190	7
0.001723	18
0.002977	52

Table P.2.6.1 Predicted Data as a
Function of Flow Rate

Q (cm^3/s)	ΔP (Pa)	$\dot{\gamma}_w$ (1/s)	τ_w (Pa)
0.24			
3.21			
5.94			
9.10			
12.39			

Table P.2.7.1 The Effect of Frying Time on the
Stress–Strain Curves of the Fried Potato Crust

| | Stress (kPa) | |
Strain	1-Min Frying Time	10-Min Frying Time
0.005	8.05	18.60
0.010	16.10	37.20
0.015	24.15	55.80
0.020	32.20	74.40
0.025	40.25	93.00

Will he be able to recover not the original but the approximate values for the missing pressure
drop, shear rate, and shear stress at the wall data according to the flow rates (Q) given in Table
P.2.6.1?

2.7. Texture is critical for the quality of french fries. In this respect, the effect of frying time on
the stress–strain curves of the fried potato crust was studied and the results are given in Table
P.2.7.1.

Calculate the modulus of elasticity for both 1-min and 10-min fried potato crust and comment
on their rigidity.

Table P.2.8.1 Stress
Relaxation Behavior of LDPE

Time (s)	Stress (Pa)
0	200.00
4	134.06
8	89.87
12	60.24
20	27.07
28	12.16
36	5.46
44	2.46
52	1.10
60	0.50

Table P.2.9.1 Creep
Compliance Data for Collagen

Time (min)	Compliance (Pa^{-1})
0.0	0
0.5	7.9×10^{-5}
1.0	13.6×10^{-5}
2.0	20.6×10^{-5}
3.0	24.1×10^{-5}
4.0	26.0×10^{-5}
5.0	26.9×10^{-5}
7.0	27.6×10^{-5}
9.0	27.8×10^{-5}

2.8. Low-density polyethylene (LDPE) is used in food packaging because of its high-impact resistance and good resistance to a wide range of chemicals. In a study, the stress relaxation
behavior of LDPE at high temperature (115°C) was examined under constant strain of 0.01
and the data are given in Table P.2.8.1.
  (a) Determine the relaxation time for LDPE. If the viscosity of LPDE at 115°C is 200 kPa · s,
    calculate its corresponding G value.
  (b) Calculate the Deborah number of the LDPE sample and compare the Deborah number of
    glass at 27°C having a relaxation time of 10^5 s for the same observation time with the LDPE.
2.9. To determine the influence of glutaraldehyde as a crosslinking agent to increase the strength of
collagen, 1% collagen solution was treated with 0.09% of glutaraldehyde. Then, the viscoelasticity of the resulting collagen gel was examined by using the creep compliance data given in
Table P.2.9.1. Examine the creep behavior of the collagen gel with the Kelvin-Voight model
and obtain the values of G, λ_{ret}, and μ.

REFERENCES

AACC (1988). *Approved Methods of the AACC,* AACC Method 74-09. St. Paul, MN: American Association of Cereal Chemists.

Abu-Jdayil, B., & Mohameed, H.A. (2004). Time-dependent flow properties of starch-milk-sugar pastes. *European Food Research & Technology, 218,* 123–127.

Arslan, E., Yener, M.E., & Esin, A. (2005). Rheological characterization of tahin/pekmez (sesame paste/concentrated grape juice) blends. *Journal of Food Engineering, 69,* 167–172.

Arslanoglu, F.N., Kar, F., & Arslan, N. (2005). Rheology of peach pulp as affected by temperature and added granular activated carbon. *Journal of Food Science and Technology-Mysore, 42,* 325–331.

Bird, R.B., Stewart, W.E., & Lightfoot, E.N. (1960). *Transport Phenomena.* New York: John Wiley & Sons.

Blahovec, J. (2003). Activation volume from stress relaxation curves in raw and cooked potato. *International Journal of Food Properties, 6,* 183–193.

Bloksma, A.H. (1990). Rheology of the breadmaking process. *Cereal Foods World, 35,* 228–236.

Bohlin, L., & Carlson, T.LG. (1980). Dynamic viscoelastic properties of wheat flour doughs: dependence on mixing time. *Cereal Chemistry, 57,* 175–181.

Bourne, M.C. (1966). Measurement of shear and compression components of puncture tests. *Journal of Food Science, 31,* 282–291.

Bourne, M.C. (1982). *Food Texture and Viscosity.* New York: Academic Press.

Bourne, M.C. (1990). Basic principles of texture measurement. In H. Faridi & J.M. Faubion (Eds.), *Dough Rheology and Baked Product Texture* (pp. 331–342). New York: AVI/Van Nostrand Reinhold.

Bruns, A.J., & Bourne, M.C. (1975). Effects of sample dimensions on the snapping force of crisp foods. Experimental verification of a mathematical model. *Journal of Texture Studies, 6,* 445–458.

Casson, N. (1959). A flow equation for pigment-oil suspensions of the printing oil type. In C.C. Mill (Ed.), *Rheology of Dispersed Systems* (pp. 82–104) New York: Pergamon Press.

Chatraei, S., Macosko, C.W., & Winter, H.H. (1981). Lubricated squeezing flow: A new biaxial extensional rheometer. *Journal of Rheology, 25,* 467–467.

Dintiz, F.R., Berhow, M.A., Bagley, E.B., Wu, Y.V., & Felker, F.C. (1996). Shear-thickening behavior and shear-induced structure in gently solubilized starches. *Cereal Chemistry, 73,* 638–643.

Dobraszczyk, B.J., & Morgenstern, M. (2003). Rheology and breadmaking process. *Journal of Cereal Science, 38,* 229–245.

Dobraszczyk, B. J., & Vincent, J. F.V. (1999). Measurement of mechanical properties of food materials in relation to texture: The materials approach. In A.J. Rosenthal (Ed.), *Food Texture: Measurement and Perception* (pp. 99–151). New York: Aspen.

Edwards, N.M., Peressini, D., Dexter, J.E., & Mulvaney, S.T. (2001). Viscoelastic properties of durum wheat and common wheat dough of different strengths. *Rheologica Acta, 40,* 142–153.

Isikli, N.D., & Karababa, E.A. (2005). Rheological characterization of fenugreek paste (cemen). *Journal of Food Engineering, 69,* 185–190.

Kaur, L., Singh, N., Sodhi, N.S., & Gujral, H.S. (2002). Some properties of potatoes and their starches. I. Cooking, textural and rheological properties of potatoes. *Food Chemistry, 79,* 177–181.

Kaya, A., &. Belibagli, K.B. (2002). Rheology of solid Gaziantep Pekmez. *Journal of Food Engineering, 54,* 221–226.

Krokida, M.K., Karathanos, V.T., & Maroulis, Z.B. (2000). Effect of osmotic dehydration on viscoelastic properties of apple and banana. *Drying Technology, 18,* 951–966.

Krokida, M.K., Maroulis, Z.B., & Saravacos, G.D. (2001). Rheological properties of fluid fruit and vegetable products: Compilation of literature data. *International Journal of Food Properties, 4,* 179–200.

Li, W., Dobraszczyk, B.J., & Schofield, J.D. (2003). Stress relaxation behavior of wheat dough, gluten and gluten fractions. *Cereal Chemistry, 80,* 333–338.

Loeb, L.B. (1965). *The Kinetic Theory of Gases,* 3rd ed. New York: Dover Publications.

McCarthy, K.L., & Seymour, J.D. (1994). Gravity current analysis of the Bostwick consistometer for power law fluids. *Journal of Texture Studies, 25,* 207–220.

Mongia, G., & Ziegler, G.R. (2000). The role of particle size distribution of suspended solids in defining the flow properties of milk chocolate. *International Journal of Food Properties, 3*, 137–147.

Mukprasirt, A., Herald, T.J., & Flores, R.A. (2000). Rheological characterization of rice flour-based batters. *Journal of Food Science, 65*, 1194–1199.

Munson, B.R., Young, D.F., & Okiishi, T.H. (1994). *Fundamentals of Fluid Mechanics*. New York: John Wiley & Sons.

Pederson, L., Kaack, K., Bergsøe, M.N., & Adler-Nissen, J. (2004). Rheological properties of biscuit dough from different cultivars, and relationship to baking characteristics. *Journal of Cereal Science, 39*, 37–46.

Pomeranz, Y. (1987). *Modern Cereal Science and Technology*. New York: VCH.

Rao, M.A. (1986). Rheological properties of fluid foods. In M.A. Rao & S.S.H. Rizvi (Eds.), *Engineering Properties of Foods* (pp. 1–48). New York: Marcel Dekker.

Rao, M A. (1999). *Rheology of Fluid and Semisolid Foods: Principles and Applications*. New York: Aspen.

Reid, R.C., Prausnitz, J.M., & Sherwood, T.K. (1977). *The Properties of Gases and Liquids*, 3rd ed. New York: McGraw-Hill.

Rha, C., & Pradipasera, P. (1986) Viscosity of proteins. In J.R. Mitchell, & D.A. Leward (Eds.), *Functional Properties of Food Macromolecules* (pp. 371–433). New York: Elsevier.

Rosell, C.M., Rojas, J.A., & de Barber, C.B. (2001). Influence of hydrocolloids on dough rheology and bread quality. *Food Hydrocolloid, 15*, 75–81.

Safari-Ardi, M., & Phan-Thien, N. (1998). Stress relaxation and oscillatory tests to distinguish between doughs prepared from wheat flours of different varietal origin. *Cereal Chemistry, 75*, 80–84.

Sakiyan, O., Sumnu, G., Sahin, S., & Bayram, G. (2004). Influence of fat content and emulsifier type on the rheological properties of cake batter. *European Food Research & Technology, 219*, 635–638.

Seyhun, N. (2002). *Retardation of Staling of Microwave Baked Cakes*. MS thesis, Middle East Technical University, Ankara, Turkey.

Simuang, J., Chiewchan, N., & Tansakul, A. (2004). Effects of fat content and temperature on apparent viscosity of coconut milk. *Journal of Food Engineering, 64*, 193–197.

Sivaramakrishnan, H.P., Senge, B., & Chattopadhyay, P.K (2004). Rheological properties of rice dough for making rice bread. *Journal of Food Engineering, 62*, 37–45.

Spies, R. (1990). Application of rheology in the bread industry. In H. Faridi & J.M. Faubion (Eds.), *Dough Rheology and Baked Product Texture* (pp. 343–361). New York: Van Nostrand Reinhold.

Steffe, J.F. (1996). *Rheological methods in Food Process Engineering,* 2nd ed. East Lansing, MI: Freeman Press (available at www.egr.msu.edu/~steffe/freebook/offer.html).

Tiziani, S., & Vodovotz, Y. (2005). Rheological effects of soy protein addition to tomato juice. *Food Hydrocolloids, 19*, 45–52.

Van Vliet, T. (1999). Rheological classification of foods and instrumental techniques for their study. In A.J. Rosenthal (Ed.), *Food Texture Measurement and Perception* (pp. 65–98). New York: Aspen.

Walstra, P. (1980). Evaluation of the firmness of butter. *International Dairy Federation Document, 135*, 4–11.

Wang, F.C., & Sun, X.S. (2002). Creep recovery of wheat flour doughs and relationship to other physical dough tests and bread making performance. *Cereal Chemistry, 79*, 567–571.

Wilkes, J.O. (1999). *Fluid Mechanics for Chemical Engineering*. Upper Saddle River, NJ: Prentice-Hall.

Zaidul, I.S.M., Karim, A.A., Manan, D.M.A., Azlan, A., Norulaini, N.A.N., & Omar, A.K.M. (2003). Stress relaxation test for sago-wheat mixtures gel. *International Journal of Food Properties, 6*, 431–442.

CHAPTER 3

Thermal Properties of Foods

SUMMARY

Since many stages in the processing and preservation of foods involve heat transfer, it is important to understand the thermal properties of foods. Thermal properties data are required in engineering and process design. An energy balance for a heating or cooling process cannot be made and the temperature profile within the material cannot be determined without knowing the thermal properties of the material. In this chapter, principles and measurement methods of thermal conductivity, specific heat, enthalpy, and thermal diffusivity are discussed. In addition, predicted models for thermal conductivity and specific heat are given.

Thermal conductivity is defined as the ability of a material to conduct heat. Thermal conductivity of foods depends on temperature, composition, and porosity of material. There are steady-state and transient-state methods for measurement of thermal conductivity. Although steady-state methods are simple in the mathematical processing of results, the long time necessary for the measurement makes transient methods more preferable for foods. The most commonly used transient methods are the thermal conductivity probe method, transient hot wire method, modified Fitch method, point heat source method, and comparative method.

Specific heat shows the amount of heat required to increase the temperature of unit mass of the substance by unit degree. There are equations in literature to express specific heat as a function of composition or temperature. Specific heat can be determined by the method of mixture, method of guarded plate, method of comparison calorimeter, adiabatic agricultural calorimeter, differential scanning calorimeter, and method of calculated specific heat. Differential scanning calorimeter can also be used to determine gelatinization enthalpy of starch samples.

Thermal diffusivity measures the ability of a material to conduct thermal energy relative to its ability to store thermal energy. Thermal diffusivity can be calculated indirectly from the measured thermal conductivity, density, and specific heat. It can also be determined directly from the solution of a one-dimensional unsteady-state heat transfer equation.

3.1 FOURIER'S LAW OF HEAT CONDUCTION

We need a driving force to overcome a resistance in order to transfer a property. For any kind of molecular transport processes (momentum, heat or thermal energy, and mass) the general equation can be written as follows:

$$\text{Rate of a transfer process} = \frac{\text{Driving force}}{\text{Resistance}}$$

(3.1)

Consider a wall of thickness X and surface area A. Imagine that the wall is initially uniform at a temperature T_0. At time $t = 0$, one side of the wall is suddenly brought to a slightly higher temperature T_1 and maintained at that temperature. Heat is conducted through the wall as a result of the temperature difference, and as time proceeds, the temperature profile in the wall changes. Finally, linear steady-state temperature distribution is achieved as shown in Fig. 3.1.

The driving force for the heat transfer to occur is the temperature difference:

$$\text{Driving force} = T_1 - T_0$$

(3.2)

While the rate of heat conduction through the wall is proportional to the heat transfer area (A), the thickness of the wall (X) provides resistance to heat transfer. In addition, the ability of the wall material to conduct heat should be considered. Each material has a different ability to conduct heat. The responses of steel and wood to heating are not the same when they are exposed to the same amount of heat. This material property is named thermal conductivity (k). Considering all these parameters, the resistance to heat transfer can be written as:

$$\text{Resistance} = \frac{X}{kA}$$

(3.3)

When the steady-state condition has been reached, the rate of heat flow (Q) through the wall can be written by substituting Eqs. (3.2) and (3.3) into Eq. (3.1):

$$Q = kA \frac{T_1 - T_0}{X}$$

(3.4)

Equation (3.4) in differential form gives Fourier's law of heat conduction:

$$Q_x = -kA \frac{dT}{dx}$$

(3.5)

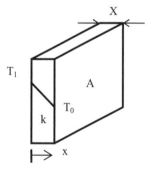

Figure 3.1 Steady-state heat transfer through the wall.

where Q_x is the rate of heat flow in the x-direction. Heat is conducted in the direction of decreasing temperature and the temperature gradient becomes negative when temperature decreases with increasing x. Therefore, a negative sign is added to Eq. (3.5).

3.2 THERMAL CONDUCTIVITY

The thermal conductivity of a material is defined as a measure of its ability to conduct heat. It has a unit of W/m K in the SI system.

A solid may be comprised of free electrons and atoms bound in a periodic arrangement called a lattice. Thermal energy is transported through the molecules as a result of two effects: lattice waves and free electrons. These two effects are additive:

$$k = k_e + k_\ell \tag{3.6}$$

In pure metals, heat conduction is based mainly on the flow of free electrons and the effect of lattice vibrations is negligible. In alloys and nonmetallic solids, which have few free electrons, heat conduction from molecule to molecule is due to lattice vibrations. Therefore, metals have higher thermal conductivities than alloys and nonmetallic solids.

The regularity of the lattice arrangement has an important effect on the lattice component of thermal conductivity. For example, diamond has very high thermal conductivity because of its well ordered structure (Fig. 3.2).

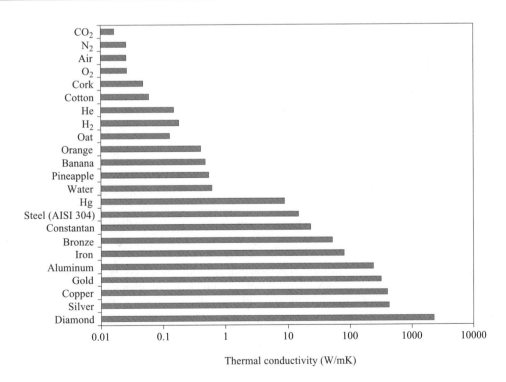

Figure 3.2   Thermal conductivities of various materials at 27°C.

As temperature increases, lattice vibrations increase. Therefore, thermal conductivities of alloys increase with an increase in temperature while the opposite trend is observed in metals because the increase in lattice vibrations impedes the motion of free electrons.

In porous solids such as foods, thermal conductivity depends mostly on composition but also on many factors that affect the heat flow paths through the material, such as void fraction, shape, size and arrangement of void spaces, the fluid contained in the pores, and homogeneity (Sweat, 1995). Thermal conductivity in foods having fibrous structures such as meat cannot be the same in different directions (anisotropy) because heat flow paths through the material change with respect to direction. Thermal conductivity increases with moisture content. Thermal conductivities of food materials vary between that of water ($k_{water} = 0.614$ W/m°C at 27°C) and that of air ($k_{air} = 0.026$ W/m°C at 27°C), which are the most and the least conductive components in foods, respectively (Fig. 3.2). The thermal conductivity values of the other food components fall between these limits. Dry porous solids are very poor heat conductors because the pores are occupied by air. For porous materials, the measured thermal conductivity is an apparent one, called the effective thermal conductivity. It is an overall thermal transport property assuming that heat is transferred by conduction through the solid and the porous phase of the material.

Thermal conductivity of ice is nearly four times greater than that of water ($k_{ice} = 2.24$ W/m°C at 0°C). This partly accounts for the difference in freezing and thawing rates of food materials. During freezing, the outer layer is in the frozen state and the ice propagates toward the center replacing nonflowing water by a better conductor. Heat is conducted through the ice layer, which has higher thermal conductivity than that of water. However, in thawing, the frozen part is at the center and heat is conducted through the thawed region, which has lower thermal conductivity or higher resistance than that of ice. As a result, the thawing process is slower than freezing process.

Above their freezing point, thermal conductivity of foods increases slightly with increasing temperature. However, around their freezing point, temperature has a strong impact on the thermal conductivity as dramatic changes in the physical nature of the foods take place during freezing or thawing.

Thermal conductivity data for more than 100 food materials in the recent literature were classified and analyzed by Krokida, Panagiotou, Maroulis, and Saravacos (2001). The thermal conductivity of different food materials was also given in Rahman (1995), Saravacos and Maroulis (2001), and Sweat (1995).

In the case of liquids and gases, heat conduction occurs as a result of molecular collisions. Since the intermolecular spacing is much larger and the motion of the molecules is more random in fluids as compared to that in solids, thermal energy transport is less effective. Therefore, thermal conductivities of fluids are lower than those of solids.

Gases have very low thermal conductivity values (Fig. 3.2). Thermal conductivities of dilute monatomic gases can be predicted by kinetic theory, which is analogous to the prediction of viscosity of gases (Loeb, 1965).

Consider a pure gas composed of rigid, nonattracting spherical molecules of diameter d and mass m present in a concentration of N molecules per unit volume. N is considered to be small enough so that the average distance between molecules is many times their diameter d. The specific heat of the gas is c_v. Assume that the gas molecules are arranged in layers normal to the z-axis and the upper layer is warmer. Then, the heat is transferred in the z-direction only. When steady state is reached, the equation of energy becomes:

$$\frac{d^2T}{dz^2} = 0 \qquad (3.7)$$

Integration states that:

$$\frac{dT}{dz} = \text{Constant} \tag{3.8}$$

Therefore, at any point in the gas:

$$T = T_0 + \frac{dT}{dz}z \tag{3.9}$$

One third of the molecules are moving along the z-axis and carry heat from one layer to the other or one third of all the velocity components lie along z.

According to kinetic theory, it is assumed that an average molecule traverses a distance equal to the mean free path between impacts. If mean free path is λ, one may consider that the length of this path is the thickness of the layer of gas. On the two sides of a gas layer having a thickness of λ, the average temperature difference of the molecules is expressed as $\lambda\frac{dT}{dz}$. Molecules coming from upper or warmer layer to the lower or cooler layer carry an excess energy of $mc_v\lambda\frac{dT}{dz}$ from upper to the lower side. It can be said that on average one third of the molecules are moving along z, and one half are moving downward. Thus, the number of molecules having average velocity of $\bar{c}$, moving downward through the layer λ, per unit area per second will be one sixth of $N\bar{c}$. The energy transferred across this layer by the molecules can be expressed as, $\frac{1}{6}N\bar{c}mc_v\lambda\frac{dT}{dz}$. Similarly $\frac{1}{6}N\bar{c}$ molecules pass upward and they carry $-\frac{1}{6}N\bar{c}mc_v\lambda\frac{dT}{dz}$ energy units across unit area per second.

The net energy transfer is the difference of energy carried up and down. Then, the energy transferred through unit area per second is:

$$q_z = -\frac{1}{3}N\bar{c}mc_v\lambda\frac{dT}{dz} \tag{3.10}$$

From Fourier's law of heat conduction:

$$q_z = -k\frac{dT}{dz} \tag{3.11}$$

Equating Eqs. (3.10) and (3.11):

$$k = \frac{1}{3}N\bar{c}mc_v\lambda \tag{3.12}$$

As discussed in Chapter 2, the mean free path is given by the following equation:

$$\lambda = \frac{1}{\sqrt{2}\pi d^2 N} \tag{2.7}$$

Inserting Eq. (2.7) into (3.12) gives:

$$k = \frac{mc_v\bar{c}}{3\sqrt{2}\pi d^2} \tag{3.13}$$

According to kinetic theory, molecular velocities relative to fluid velocity have an average magnitude given by the following equation:

$$\bar{c} = \sqrt{\frac{8RT}{\pi N_A m}} \tag{2.9}$$

where N_A is the Avogadro number, m is the mass of the molecule, R is the gas constant, and T is the absolute temperature. Inserting Eq. (2.9) into Eq. (3.13) gives:

$$k = \frac{2c_v}{3\pi^{3/2}d^2}\sqrt{mKT}$$ (3.14)

where $K = R/N_A$ and it is the Boltzmann constant.

This equation predicts that thermal conductivity should increase with the square root of temperature. Thermal conductivities of gases are independent of pressure over a wide range.

The mechanism of heat conduction in liquids is qualitatively the same as in gases. However, the situation is more complicated since the molecules are more closely spaced and they exert a stronger intermolecular force field. The thermal conductivities of liquids usually lie between those of solids and gases. Unlike those of gases, thermal conductivities of nonmetallic liquids usually decrease as temperature increases. There are some exceptions to this behavior such as water and glycerine. Polar or associated liquids such as water may exhibit a maximum in the thermal conductivity versus temperature curve. Generally, thermal conductivity decreases with increasing molecular weight.

3.2.1 Prediction of Thermal Conductivity

Predictive models have been used to estimate the effective thermal conductivity of foods. A number of models for thermal conductivity exist in the literature (Murakami & Okos, 1989); however many of them contain empirical factors and product-specific information, which restrict their applicability.

Modeling of thermal conductivity based on composition has been a subject of considerable interest. It is important to include the effect of air in the porous foods and ice in the case of frozen foods.

Temperature dependence of thermal conductivities of major food components has been studied. Thermal conductivities of pure water, carbohydrate (CHO), protein, fat, ash, and ice at different temperatures can be empirically expressed according to Choi and Okos (1986) as follows:

$$k_{\text{water}} = 0.57109 + 1.7625 \times 10^{-3}T - 6.7036 \times 10^{-6}T^2$$ (3.15)

$$k_{\text{CHO}} = 0.20141 + 1.3874 \times 10^{-3}T - 4.3312 \times 10^{-6}T^2$$ (3.16)

$$k_{\text{protein}} = 0.17881 + 1.1958 \times 10^{-3}T - 2.7178 \times 10^{-6}T^2$$ (3.17)

$$k_{\text{fat}} = 0.18071 - 2.7604 \times 10^{-3}T - 1.7749 \times 10^{-7}T^2$$ (3.18)

$$k_{\text{ash}} = 0.32961 + 1.4011 \times 10^{-3}T - 2.9069 \times 10^{-6}T^2$$ (3.19)

$$k_{\text{ice}} = 2.2196 - 6.2489 \times 10^{-3}T + 1.0154 \times 10^{-4}T^2$$ (3.20)

where thermal conductivities (k) are in W/m°C; temperature (T) is in °C and varies between 0 and 90°C in these equations.

Rahman (1995) correlated thermal conductivity of moist air at different temperatures from the data of Luikov (1964) as:

$$k_{\text{air}} = 0.0076 + 7.85 \times 10^{-4}T + 0.0156RH$$ (3.21)

where RH is the relative humidity, changing from 0 to 1, and temperature varies from 20 to 60°C.

Generally, in multiphase systems, the effect of geometric distribution of phases is taken into account by using structural models. There are also some models that assume an isotropic physical structure. The most well known models in literature are parallel and series models; the Krischer model, which is obtained by combining the parallel and series models; and the Maxwell-Eucken and Kopelman models used for two-component food systems consisting of a continuous and a dispersed phase. Several other models have been reviewed by Rahman (1995).

3.2.1.1   Parallel Model

In the parallel model, components are assumed to be placed parallel to the direction of heat flow (Fig. 3.3a). The effective thermal conductivity of a food material made of n components can be calculated using volume fractions (X_i^v) and thermal conductivities (k_i) of each component (i) from the following equation:

$$k_{pa} = \sum_{i=1}^{n} k_i X_i^v \tag{3.22}$$

where

$$X_i^v = \frac{X_i^w / \rho_i}{\sum\limits_{i=1}^{n} \left(X_i^w / \rho_i \right)} \tag{3.23}$$

where

   X_i^v = volume fraction of the ith constituent,
   X_i^w = mass fraction of the ith constituent,
   ρ_i = density of the ith constituent (kg/m^3).

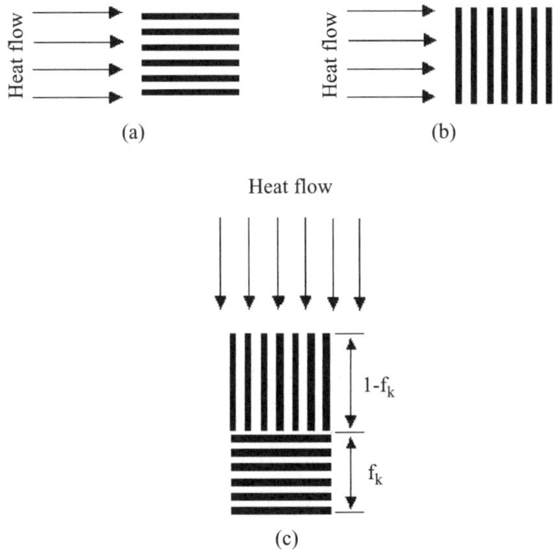

Figure 3.3   (a) Parallel, (b) Series, and (c) Krischer models.

The parallel distribution results in maximum thermal conductivity value. If the food material is assumed to be composed of three components (water, solid, and air), effective thermal conductivity can be calculated from:

$$k_{pa} = k_w X_w^v + k_s X_s^v + k_a X_a^v \qquad (3.24)$$

where X_w^v, X_s^v, and X_a^v are the volume fractions of moisture, solid, and air, respectively and k_w, k_s, and k_a are the corresponding thermal conductivities.

3.2.1.2 Series (Perpendicular) Model

In the perpendicular model, components are assumed to be placed perpendicular to the direction of heat flow (Fig. 3.3b). The effective thermal conductivity of a food material can be calculated from the following equation:

$$\frac{1}{k_{se}} = \sum_{i=1}^{n} \frac{X_i^v}{k_i} \qquad (3.25)$$

The perpendicular distribution results in a minimum thermal conductivity value. If the food material is assumed to be composed of three components (water, solid, and air), effective thermal conductivity can be calculated from:

$$\frac{1}{k_{se}} = \frac{X_w^v}{k_w} + \frac{X_s^v}{k_s} + \frac{X_a^v}{k_a} \qquad (3.26)$$

3.2.1.3 Krischer Model

The real value of thermal conductivity should be somewhat between the thermal conductivities derived from the parallel model and series model. The series and parallel models do not take into account the natural distribution of component phases. Thus, Krischer proposed a generalized model by combining the parallel and series models using a phase distribution factor. The distribution factor, f_k, is a weighing factor between these extreme cases (Fig. 3.3c). Krischer's model is described by the following equation:

$$k = \frac{1}{\dfrac{1 - f_k}{k_{pa}} + \dfrac{f_k}{k_{se}}} \qquad (3.27)$$

where k, k_{pa}, and k_{se} are the effective thermal conductivity by the Krischer, parallel, and series models, respectively.

The disadvantage of this model is that it is not possible to find the values of f_k without making experiments because f_k depends on moisture content, porosity, and temperature of foods.

3.2.1.4 Maxwell-Eucken Model

Krischer's model is one of the most widely used one in food engineering literature. However, for isotropic materials, it is not very useful.

For a two-component food system consisting of a continuous and a dispersed phase, a formula based on the dielectric theory had been developed by Maxwell (Maxwell, 1904). The model was then adapted by Eucken (Eucken, 1940) and is known as the Maxwell-Eucken model. The effective thermal conductivity of a food by this model is defined as:

$$k = k_c \left(\frac{2k_c + k_d - 2X_d^v(k_c - k_d)}{2k_c + k_d + X_d^v(k_c - k_d)} \right)$$ (3.28)

where k_c and k_d are the thermal conductivities of continuous and dispersed phases, respectively.

A three-step modification of the Maxwell-Eucken model was developed for porous foods by Hamdami, Monteau, and Le Bail (2003). In the first step, water was the continuous and ice was the discontinuous phase. In the second step, the solid was the continuous and water-ice was the discontinuous phase. In the third step, solid-water-ice was the continuous and air the discontinuous phase.

Carson, Lovatt, Tanner, and Cleland (2005) recommended a modified version of the Maxwell-Eucken model. They proposed that porous materials should be divided into two types according to porosity (external or internal porosity) because the mechanism of heat conduction in these two types differs.

3.2.1.5 Kopelman Model

The isotropic model of Kopelman (1966) describes the thermal conductivity of a composite material as a combination of continuous and discontinuous phases.

$$k = \frac{k_c [1 - Q]}{1 - Q \left[1 - (X_d^v)^{1/3} \right]}$$ (3.29)

where

$$Q = (X_d^v)^{2/3} \left(1 - \frac{k_d}{k_c} \right)$$ (3.30)

This equation is useful for two-component models, but needs modification for multicomponent systems. In particular, the definition of continuous and dispersed phases needs specification. Since foods are multicomponent systems, and the phases are associated with each other in complex ways, a stepwise procedure in which the mean thermal conductivity of pairs of components is found at each step in the Kopelman and Maxwell equations (Hamdami et al., 2003).

Moreira, Palau, Sweat, and Sun (1995) and Sahin, Sastry, and Bayındırlı (1999) used the isotropic model of Kopelman to predict the effective thermal conductivity of tortilla chips and french fries, respectively. The approach chosen in Kopelman equation is to successively determine the thermal conductivity of two-component systems, starting with water continuous and carbohydrate dispersed phases. Then, the water–carbohydrate phase is taken as continuous and protein as dispersed phases and the iterative procedure is continued through all phases using the order: water (phase 1), carbohydrate (2), protein (3), fat (4), ice (5), ash (6), and air (7). The following iterative algorithm was obtained for the thermal conductivity of a system of $i + 1$ components:

$$k_{comp,i+1} = \frac{k_i [1 - Q_{i+1}]}{1 - Q_{i+1} \left[1 - (X_{d,i+1}^v)^{1/3} \right]}$$ (3.31)

where the following definitions apply:

$$Q_{i+1} = (X_{d,i+1}^v)^{2/3} \left(1 - \frac{k_{i+1}}{k_i} \right)$$ (3.32)

$$X_{d,i+1}^v = \frac{V_{i+1}}{\sum\limits_{i}^{i+1} V_i}$$ (3.33)

3.2.1.6 Improved Thermal Conductivity Prediction Models

Many food products are porous and the thermal conductivity strongly depends on porosity since thermal conductivity of air in pores is an order of magnitude lower than the thermal conductivities of other food components.

In general, the effective thermal conductivity of heterogeneous materials cannot be predicted with simple additive models. Heat is transported in pores of food materials not only by conduction but also by latent heat. Therefore, for porous food materials, the two kinds of heat transport should be considered in the modeling of effective thermal conductivity. Effective thermal conductivity of a highly porous model food, at above and below freezing temperatures, was studied by Hamdami et al. (2003). The Krischer model was in good agreement with the experimental data. The model, including the effect of evaporation–condensation phenomena in addition to heat conduction, was also useful in predicting the effective thermal conductivity of porous model food. If there is heat transfer in porous food material, moisture migrates as vapor in the pore space as a result of the vapor pressure gradient caused by the temperature gradient. The effective thermal conductivity in pores can be expressed by the following equation by considering the effect of latent heat transport:

$$k_{\text{pores}} = k_{\text{air}} + k_{\text{eva-con}} f_{\text{eva-con}} \tag{3.34}$$

where

k_{pores} = effective thermal conductivity in pores (W/m K),
k_{air} = thermal conductivity of air (W/m K),
$k_{\text{eva-con}}$ = equivalent thermal conductivity due to evaporation–condensation phenomena (W/m K),
$f_{\text{eva-con}}$ = resistance factor against vapor transport.

Equivalent thermal conductivity due to evaporation–condensation phenomena ($k_{\text{eva-con}}$) can be calculated from:

$$k_{\text{eva-con}} = \frac{D}{RT} \frac{P}{P - a_w P_{\text{sat}}} \lambda a_w \frac{dP_{\text{sat}}}{dT} \tag{3.35}$$

where

D = diffusivity of water vapor in air (m^2/s),
R = gas constant (8.3145 J/mol K),
T = temperature (K),
P = total pressure (Pa),
P_{sat} = saturated pressure of water vapor (Pa),
a_w = water activity,
λ = latent heat of evaporation (J/mol).

Thermal conductivities of foods change during processing with variation in composition. For example, thermal conductivity decreases with decreasing moisture content during drying. The formation of pores (air phase) in foods further decreases the thermal conductivity. Rahman (1992) developed a thermal conductivity model for beef, apple, potato, pear, and squid during drying by introducing a porosity term:

$$\left(\frac{k}{k_0}\right)\left(\frac{1}{1-\varepsilon}\right) = \left[1.82 - 1.66 \exp\left(-0.85 \frac{X_w^w}{X_{w0}^w}\right)\right] \tag{3.36}$$

where

k = effective thermal conductivity (W/m K),
k_0 = initial thermal conductivity (W/m K),
ε = volume fraction of air or porosity,
X_w^w = mass fraction of water (wet basis),
X_{w0}^w = mass fraction of initial water (wet basis).

The above correlation is an extension of the parallel model, which is realistic in the case of homogeneous food materials. It is valid for moisture content from 5% to 88% (wet basis), porosity from 0 to 0.5, and temperature from 20 to 25°C. However, this correlation has some disadvantages: When porosity is 1.0, the left side of the correlation becomes infinity, which is physically incorrect (Rahman & Chen, 1995). Moreover, the effect of temperature on thermal conductivity is not taken into account. In addition, the thermal conductivity values of fresh foods (i.e., before processing) are required. Therefore, an improved general thermal conductivity prediction model has been developed for fruits and vegetables during drying as a function of moisture content, porosity, and temperature (Rahman, Chen, & Perera, 1997):

$$\frac{\varphi}{1 - \varepsilon + \left[k_{\text{air}} / (k_w)_{\text{ref}} \right]} = 0.996 \left(T / T_{\text{ref}} \right)^{0.713} \left(X_w^w \right)^{0.285} \tag{3.37}$$

where T_{ref} is the reference temperature which is 0°C and $(k_w)_{\text{ref}}$ is the thermal conductivity of water evaluated at reference temperature. φ is the Rahman-Chen structural factor related to the phase distribution of components of food material and expressed as a function of temperature, moisture content, and porosity (Rahman, 1995) as:

$$\varphi = \frac{k - \varepsilon k_{\text{air}}}{(1 - \varepsilon - X_w^w) k_s + X_w^w k_w} \tag{3.38}$$

In the development of this model, the moisture content varied from 14% to 88% (wet basis), porosity varied from 0 to 0.56, and temperature varied from 5 to 100°C.

Empirical models, relating thermal conductivity to temperature and moisture content of the material, are fitted to literature data for different food materials (Maroulis, Saravacos, Krokida, & Panagiotou, 2002). The thermal conductivities of the materials with intermediate moisture were estimated using the two-phase parallel model assuming that the material is a uniform mixture of two components: a dried material and a wet material with infinite moisture.

$$k = (1 - X_w^w) k_{\text{dry}} + X_w^w k_{\text{wet}} \tag{3.39}$$

where

k = effective thermal conductivity (W/m K),
k_{dry} = thermal conductivity of the dried material (phase a) (W/m K),
k_{wet} = thermal conductivity of the wet material (phase b) (W/m K),
X_w^w = mass fraction of moisture in the material (w/w in db).

The temperature dependence of the thermal conductivities of both phases is expressed by an Arrhenius type equation:

$$k_{\text{dry}} = k_0 \exp \left[-\frac{E_0}{R} \left(\frac{1}{T} - \frac{1}{T_{\text{ref}}} \right) \right] \tag{3.40}$$

$$k_{\text{wet}} = k_i \exp \left[-\frac{E_i}{R} \left(\frac{1}{T} - \frac{1}{T_{\text{ref}}} \right) \right] \tag{3.41}$$

Table E.3.1.1. Composition of the Date Fruit and Densities of Food Components at 25°C

Component	Weight (%)	Density (kg/m³)
Water	22.5	995.7
Carbohydrate	72.9	1592.9
Protein	2.2	1319.6
Fat	0.5	917.15
Ash	1.9	2418.2

where

T_{ref} = reference temperature (60°C) which is a typical temperature of air drying of foods,

R = the ideal gas constant (8.3143 J/mol K),

k_0 = thermal conductivity at moisture $X = 0$ and temperature $T = T_{ref}$ (W/m K),

k_i = thermal conductivity at moisture $X = \infty$ and temperature $T = T_{ref}$ (W/m K),

E_0 = activation energy for heat conduction in dry material at $X = 0$ (J/mol),

E_i = activation energy for heat conduction in wet material at $X = \infty$ (J/mol).

Example 3.1. The composition of date fruit (*Phoenix dactylifera*) and the densities of food components are given in Table E.3.1.1. Determine the thermal conductivity of the fruit at 25°C, using parallel, series, and isotropic Kopelman models.

Solution:

To calculate the thermal conductivity of the fruit using predictive models, thermal conductivity values of food components at 25°C are required. They can be calculated using Eqs. (3.15)–(3.19) (Table E.3.1.2).

Using the composition and density of components data given in the question, the specific volume of each component is calculated:

$$\text{Specific volume of component } i = \frac{\text{Mass fraction of component } i}{\text{Density of component } i}$$

Then, total specific volume is determined by adding the volume of each component and found as 7.14×10^{-4} m³. Volume fractions of components are calculated by dividing the component volume to total volume. Specific volumes and volume fractions of each component are given in Table E.3.1.3.

Table E.3.1.2. Thermal Conductivity Values of Components at 25°C

Component	Thermal Conductivity Equation	Eq. no.	k_i (W/m K)
Water	$k_{water} = 0.57109 + 1.7625 \times 10^{-3} T - 6.7036 \times 10^{-6} T^2$	(3.15)	0.610
Carbohydrate	$k_{CHO} = 0.20141 + 1.3874 \times 10^{-3} T - 4.3312 \times 10^{-6} T^2$	(3.16)	0.233
Protein	$k_{protein} = 0.17881 + 1.1958 \times 10^{-3} T - 2.7178 \times 10^{-6} T^2$	(3.17)	0.207
Fat	$k_{fat} = 0.18071 - 2.7604 \times 10^{-3} T - 1.7749 \times 10^{-7} T^2$	(3.18)	0.112
Ash	$k_{ash} = 0.32961 + 1.4011 \times 10^{-3} T - -2.9069 \times 10^{-6} T^2$	(3.19)	0.363

Table E.3.1.3. Specific Volume and Volume Fraction of Components in Date Fruit

Component	Specific Volume (m^3/kg)	Volume Fraction of Components (X_i^v)
Water	2.26×10^{-4}	0.320
Carbohydrate	4.58×10^{-4}	0.640
Protein	1.67×10^{-5}	0.023
Fat	5.45×10^{-6}	0.0076
Ash	7.86×10^{-6}	0.011

Thermal conductivity of date fruit is calculated by using volume fractions and thermal conductivity values of components using the parallel model (Eq. 3.22):

$$k_{pa} = \sum_{i=1}^{n} k_i X_i^v \tag{3.22}$$

$$k_{pa} = (0.61)(0.32) + (0.233)(0.64) + (0.207)(0.023)$$

$$+ (0.112)(0.0076) + (0.363)(0.011) = 0.353 \text{ W/mK}$$

Using the series model (Eq. 3.25):

$$\frac{1}{k_{se}} = \sum_{i=1}^{n} \frac{X_i^v}{k_i} \tag{3.25}$$

$$= \frac{0.32}{0.61} + \frac{0.64}{0.233} + \frac{0.023}{0.207} + \frac{0.0076}{0.112} + \frac{0.011}{0.363}$$

$$= 3.48 \text{ mK/W}$$

$$k_{se} = \frac{1}{3.48} = 0.287 \text{ W/mK}$$

In the case of the isotropic model of Kopelman (Eqs. 3.31–3.33), the k value of date fruit is calculated using the order of water (1), carbohydrate (2), protein (3), fat (4), and ash (5) in the iteration.

$$k_{comp,i+1} = \frac{k_i(1 - Q_{i+1})}{1 - Q_{i+1}\left[1 - \left(X_{d,i+1}^v\right)^{1/3}\right]} \tag{3.31}$$

where

$$Q_{i+1} = \left(X_{d,i+1}^v\right)^{2/3}\left[1 - \frac{k_{i+1}}{k_i}\right] \tag{3.32}$$

$$X_{d,i+1}^v = \frac{V_{i+1}}{\sum_{i}^{i+1} V_i} \tag{3.33}$$

Starting with the water continuous and carbohydrate (CHO) dispersed phase, the thermal conductivity of water–CHO system is calculated as:

$$X_{d,\text{CHO}}^v = \frac{V_{\text{CHO}}}{V_{\text{water}} + V_{\text{CHO}}} = \frac{4.58 \times 10^{-4}}{2.26 \times 10^{-4} + 4.58 \times 10^{-4}} = 0.669$$

$$Q_{\text{CHO}} = \left(X_{d,\text{CHO}}^v\right)^{2/3}\left[1 - \frac{k_{\text{CHO}}}{k_{\text{water}}}\right]$$

$$Q_{\text{CHO}} = (0.669)^{2/3} \left[1 - \frac{0.233}{0.610} \right] = 0.473$$

$$k_{\text{water-CHO}} = \frac{k_{\text{water}}(1 - Q_{\text{CHO}})}{1 - Q_{\text{CHO}} \left[1 - \left(X_{d,\text{CHO}}^{v} \right)^{1/3} \right]}$$

$$k_{\text{water-CHO}} = \frac{0.61(1 - 0.473)}{1 - 0.473 \left[1 - (0.669)^{1/3} \right]} = 0.342 \text{ W/m K}$$

Then, the water–carbohydrate phase is taken as the continuous phase and protein as the dispersed phase.

$$X_{d,\text{prot}}^{v} = \frac{V_{\text{prot}}}{V_{\text{water}} + V_{\text{CHO}} + V_{\text{prot}}} = \frac{1.67 \times 10^{-5}}{2.26 \times 10^{-4} + 4.58 \times 10^{-4} + 1.67 \times 10^{-5}} = 0.024$$

$$Q_{\text{prot}} = \left(X_{d,\text{prot}}^{v} \right)^{2/3} \left[1 - \frac{k_{\text{prot}}}{k_{\text{water-CHO}}} \right]$$

$$Q_{\text{prot}} = (0.024)^{2/3} \left[1 - \frac{0.207}{0.342} \right] = 0.033$$

$$k_{\text{water-CHO-prot}} = \frac{k_{\text{water-CHO}}(1 - Q_{\text{prot}})}{1 - Q_{\text{prot}} \left[1 - \left(X_{d,\text{prot}}^{v} \right)^{1/3} \right]}$$

$$k_{\text{water-CHO-prot}} = \frac{0.342(1 - 0.033)}{1 - 0.033 \left[1 - (0.024)^{1/3} \right]} = 0.338 \text{ W/m K}$$

The same procedure is followed throughout all phases to find the thermal conductivity of date fruit. The calculated values in the iterative procedure are shown in Table E.3.1.4. The thermal conductivity of date fruit is found to be 0.337 W/m K using the isotropic model of Kopelman.

3.2.2   Measurement of Thermal Conductivity

Measurement of thermal conductivity can be done by either steady-state or transient-state methods. There are a number of experimental measurement techniques under each of these two categories. Mohsenin (1980) and Rahman (1995) previously reviewed various thermal properties measurement methods for food materials.

Table E.3.1.4.   Results Obtained in the Iterative Procedure

System	X_d^v	Q	k (W/m K)
Water–CHO	0.669	0.473	0.342
Water–CHO–protein	0.024	0.033	0.338
Water–CHO–protein–fat	0.008	0.026	0.337
Water–CHO–protein–fat–ash (date fruit)	0.011	−0.004	0.337

The advantages of steady-state methods are the simplicity in the mathematical processing of the results, the ease of control of the experimental conditions, and often quite high precision in the results. However, a long time is required for temperature equilibration. The moisture migration and the necessity to prevent heat losses to the environment during this long measurement time are the disadvantages of steady-state methods (Ohlsson, 1983). In addition, these methods require definite geometry of the sample and relatively large sample size.

On the other hand, the transient methods are faster and more versatile than the steady-state methods and are preferable for extensive experimental measurements. Transient methods are preferred over steady-state methods because of the short experimental duration and minimization of moisture migration problems.

3.2.2.1 *Steady State Methods*

In steady-state methods, two sides of a flat object are maintained at constant but different temperatures and the heat flux through the sample is measured. Steady-state methods are longitudinal heat flow, radial heat flow, heat of vaporization, heat flux, and differential scanning calorimeter methods.

(a)   Longitudinal Heat Flow Method. The most common method in this group is guarded hot plate method. This method is suitable mostly for determination of thermal conductivity of dry homogeneous materials in slab forms. It is the most widely used and the most accurate method for the measurement of thermal conductivity of materials, which are poor heat conductors.

In this method, the heat source (T_1), the sample, and the heat sink (T_2) are placed in contact with each other and with a thermal guard heated electrically (Mohsenin, 1980). The thermal guard plates are kept at the same temperature as the adjacent surfaces, in a way that no heat leakage takes place from source, sample, or sink boundaries. Thermal conductivity is measured after the sample has reached steady-state condition. However, achieving steady-state conditions may take several hours. It is assumed that all of the measured heat input is transferred across the sample. The thermal conductivity is calculated by measuring the amount of heat input required to maintain the unidirectional steady-state temperature profile across the test sample.

$$k = \frac{QL}{A\,(T_1 - T_2)} \tag{3.42}$$

where

  k = thermal conductivity of sample (W/m K),
  Q = measured rate of heat input (W),
  L = sample thickness (m),
  A = area of sample (m^2).

(b)   Radial Heat Flow Methods. These methods are suitable mostly for loose, powdered, or granular materials.

(i)   Concentric Cylinder Method. In this method, the sample is placed between two concentric cylinders. This method is preferable for liquid samples. The heater is usually located at the outer cylinder. A coolant fluid flows through the inner cylinder. The heat that the coolant absorbed is assumed to be equal to the heat transferred through the sample. Thermal conductivity can be calculated from the

unidirectional radial steady-state heat transfer equation as:

$$k = \frac{Q \ln(r_2/r_1)}{2\pi L(T_1 - T_2)} \tag{3.43}$$

where

Q = power used by central heater (W),
L = length of the cylinder (m),
T_1 = temperature of the sample at the outer surface of radius r_1 (K),
T_2 = temperature of the sample at the inner surface of radius r_2 (K).

The length-to-diameter ratio of the cylinder must allow the radial heat flow assumption. End guard heaters may be used to minimize the error due to axial heat flow.

(ii) Concentric Cylinder Comparative Method. This method uses a central heater followed by a cylindrical sample and a cylindrical standard. The temperatures T_1 and T_2 at radii r_1 and r_2 of the sample, respectively, and temperature T_3 and T_4 at radii r_3 and r_4 of the standard, respectively, are measured. Assuming radial heat flow, the thermal conductivity can be determined from Eq. (3.44):

$$k = \frac{k_{\text{ref}}(T_3 - T_4)\ln(r_2/r_1)}{(T_1 - T_2)\ln(r_4/r_3)} \tag{3.44}$$

where k_{ref} is the thermal conductivity of the standard.

(iii) Sphere with Central Heating Source. In this method, the sample is placed between the central heater which has a radius r_1 and the outer radius of sphere, r_2. The sample completely encloses the heating source so that end losses are eliminated. Assuming that the inner and outer surfaces of the sample are T_1 and T_2, respectively, after the steady state has been established heat flow will essentially be radial and Eq. (3.45) can be used to determine thermal conductivity:

$$k = \frac{Q(1/r_1 - 1/r_2)}{4\pi(T_1 - T_2)} \tag{3.45}$$

This is the most sensitive method among the steady-state methods because the error due to heat losses can be practically eliminated. However, it cannot be widely used because of the difficulty in obtaining suitably shaped food samples. This method has been used mainly for granular materials. Samples should be filled in a vacuum environment because air bubbles trapped inside the sphere could increase contact resistance.

(c)   Heat of Vaporization Method. In this method, a small test sample is put between two silver plates, one of which is in contact with a liquid A at its boiling point and the other one is in contact with liquid B (Mohsenin, 1980). Heat transferred through the sample vaporizes some of the liquid B, which has a lower boiling point. Since the time necessary to vaporize a unit mass of liquid B is known, the thermal conductivity of the sample is calculated using Eq. (3.46):

$$k = \frac{\lambda L}{\theta A(T_A - T_B)} \tag{3.46}$$

where
θ = time necessary to vaporize a unit mass of liquid B (s/kg),
λ = heat of vaporization of the liquid at lower boiling point (J/kg),

T_A, T_B = boiling points of liquids A and B, respectively (K),
L = thickness of sample (m),
A = area of sample (m^2).

(d) Heat Flux Method. The heat flow meter is a device for measuring heat flux. It is suitable for materials with conductance (k/L) less than 11.3 W/m^{2}K (Haas & Felsenstein, 1978).

Tong and Sheen (1992) proposed a technique based on heat flux to determine effective thermal conductivity of multilayered constructions. In this technique, a heat flux sensor is attached to the inner surface of the wall with a very thin layer of high thermal conductivity adhesive. A temperature difference of 5 to 7°C is maintained within the system and the thermal conductivity is evaluated at the arithmetic mean temperature. At steady state, the heat flux is:

$$q = U\Delta T \tag{3.47}$$

where q is the heat flux in W/m^2 and U is the overall heat transfer coefficient in W/m^{2}K.

The overall heat transfer coefficient can be written in terms of convective and conductive resistances:

$$\frac{1}{U} = \frac{1}{h_i} + \frac{1}{h_o} + \sum_{i=1}^{N} \frac{L_i}{k_i} \tag{3.48}$$

where

h_i = internal heat transfer coefficient (W/m^{2}K),
h_o = external heat transfer coefficient (W/m^{2}K),
N = number of layers,
L_i = thickness of layer i (m),
k_i = thermal conductivity of layer i (W/mK).

If h_i and h_o are very large, convective resistances are negligible and Eq. (3.48) becomes:

$$\frac{1}{U} = \sum_{i=1}^{N} \frac{L_i}{k_i} \tag{3.49}$$

Since it is not possible to measure the thermal conductivity of each layer, an effective thermal conductivity (k_{eff}) can be used:

$$\sum_{i=1}^{N} \frac{L_i}{k_i} = \frac{\sum_{i=1}^{N} L_i}{k_{\text{eff}}} \tag{3.50}$$

Combining Eqs. (3.49) and (3.50) gives:

$$k_{\text{eff}} = U \sum_{i=1}^{N} L_i \tag{3.51}$$

Then, substituting Eq. (3.47) into Eq. (3.51):

$$k_{\text{eff}} = \frac{q \sum_{i=1}^{N} L_i}{\Delta T} \tag{3.52}$$

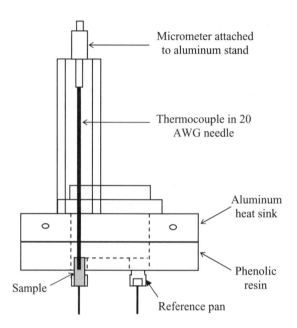

Figure 3.4 Illustration of sample placement in DSC method for determination of thermal conductivity. [From Buhri, A.B. & Singh, R.P. Measurement of food thermal conductivity using differential scanning calorimetry. *Journal of Food Science, 58*(5), 1145–1147. Copyright © (1993) with permission from IFT.]

(e)   Differential Scanning Calorimeter (DSC). An attachment to a differential scanning calorimeter was designed to measure thermal conductivity of foods as shown in Fig. 3.4 (Buhri & Singh, 1993). The sample of uniform cross section (possibly cylindrical) is placed in the sample pan, the opposite end of which is in contact with a heat sink at constant temperature. Initially, the sample is maintained at a constant temperature. At a predetermined time, the pan temperature is immediately increased to a predetermined higher value. A new steady state is reached in a few minutes and the heat flow into the DSC pan levels off. A typical DSC response curve is shown in Fig. 3.5. The difference in heat flow (ΔQ) between the two states is recorded from the thermogram. Then, thermal conductivity of the sample can be calculated using Fourier's heat conduction equation:

$$k = \frac{L\Delta Q}{A(\Delta T_2 - \Delta T_1)} \tag{3.53}$$

where L is the sample thickness, A is the cross-sectional area, ΔT_1 is the initial temperature difference, and ΔT_2 is the final temperature difference.

This method is simple and suitable for small size samples, for both low- and high-moisture foods. Time to achieve the new steady state is small enough (10 to 15 min) to prevent moisture migration since the sample is small.

This approach may be modified to measure thermal conductivity as a function of temperature by using small thermal perturbations (Sastry & Cornelius, 2002). However, measurement of thermal conductivity under ultrahigh temperature (UHT) conditions may require extensive equipment

Heat Flow

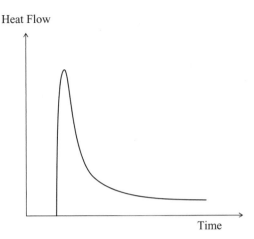

Time

Figure 3.5 Typical DSC thermogram for measurement of thermal conductivity [From Buhri, A.B. & Singh, R.P. Measurement of food thermal conductivity using differential scanning calorimetry. *Journal of Food Science, 58*(5), 1145–1147. Copyright © (1993) with permission from IFT.]

modification. Thus, it is more expensive than either the line heat source or modified Fitch methods, which are the most commonly used unsteady-state thermal conductivity measurement methods in food systems.

3.2.2.2 Unsteady-State Methods

The most important transient methods are the thermal conductivity probe method, transient hot wire method, modified Fitch method, point heat source method, and comparative method.

(a) Thermal Conductivity Probe Method. The theory of the thermal conductivity probe or line heat source method has been reviewed by many authors (Hooper & Lepper, 1950; Murakami, Sweat, Sastry, & Kolbe, 1996a; Nix, Lowery, Vachon, & Tanger, 1967). This method is the most popular method for determining thermal conductivity of food materials because of its relative simplicity and speed of measurement. In addition, this method requires relatively small sample sizes. On the other hand, it requires a fairly sophisticated data acquisition system (Sweat, 1995).

In this method, a constant heat source is applied to an infinite solid along a line with infinitesimal diameter, such as a thin resistant wire. The electrical wire must have a low resistance so that the voltage drop across it is negligible compared to the voltage drop across the heater. The cross section of the line heat source probe and the experimental apparatus for measurement of the effective thermal conductivity using the line heat probe are shown in Figs. 3.6 and 3.7, respectively.

For measurement of thermal conductivity, the container is filled with sample and the line heat source probe is inserted at the center of the container. The container is placed in a constant temperature bath and equilibrated at room temperature. After the initial temperature is recorded, the probe heater is activated and heated at a constant rate of energy input. Then, the time versus temperature adjacent to the line heat source is recorded.

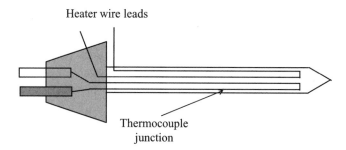

Figure 3.6   Cross section of thermal conductivity probe. [From Sastry, S.K., & Cornelius, B.D. (2002). *Aseptic Processing of Foods Containing Solid Particulates.* New York: John Wiley & Sons. Copyright © (2002) with permission from John Wiley.]

Theoretically, the line heat source technique is based on unsteady state heat conduction in an infinite medium and is expressed by:

$$\frac{\partial T}{\partial t} = \alpha \, \nabla^2 T \tag{3.54}$$

where α is the thermal diffusivity of the material $\left(\alpha = {}^{k}\!/_{\rho \, c_p}\right)$.

The initial condition is:

$$\text{at } t = 0 \quad \Delta T(r, t) = 0 \tag{3.55}$$

The line heat source theory is based on a line heat source of infinite length with negligible axial heat flow. Thermal conductivity is measured only in a direction radial to the probe because

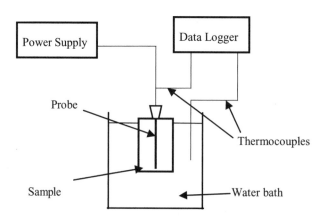

Figure 3.7   Experimental apparatus for measurement of the effective thermal conductivity using line heat probe. [From Sahin, S., Sastry, S.K., & Bayındırlı, L. Effective thermal conductivity of potato during frying: Measurement and modelling. *International Journal of Food Properties, 2*, 151–161. Copyright © (1999) with permission from Taylor & Francis.]

the heat flow is only in the radial direction. Therefore, the following boundary conditions can be written:

$$\text{B.C.1 at } r = 0 \left(r \frac{\partial T}{\partial r} \right) = -\frac{Q}{2\pi k} \tag{3.56}$$

$$\text{B.C.2 at } r = \infty \quad \Delta T(r, t) = 0 \tag{3.57}$$

It is required to assume that the power input per unit probe length, Q and the thermal properties are constant and the thermal mass of the heater is negligible. The solution of Eq. (3.54) is (Carslaw & Jaeger, 1959):

$$\Delta T = \frac{Q}{4\pi k} \int_{\beta^2}^{\infty} \frac{e^{-u}}{u} du \tag{3.58}$$

where

$$u = \frac{r^2}{4\alpha (t - t_0)} \tag{3.59}$$

$$\beta = \frac{r}{2\sqrt{\alpha t}} \tag{3.60}$$

The analytical solution of Eq. (3.58) gives:

$$\Delta T = -\frac{Q}{4\pi k} E_1 \left[-\beta^2 \right] \tag{3.61}$$

where E_1 is the first-order exponential integral function, which can be evaluated using the following equation (Abramowitz & Stegun, 1964):

$$E_1 \left[-\beta^2 \right] = C_e + \ln(\beta^2) + \sum_{n=1}^{\infty} \left[\frac{(-1)^n \beta^{2n}}{nn!} \right] \tag{3.62}$$

The Euler constant, C_e, is equal to 0.57721. Then, Eq. (3.61) can be written as:

$$\Delta T = \frac{Q}{4\pi k} \left[-0.57721 - \ln(\beta^2) + \frac{\beta^2}{2!} - \frac{\beta^4}{4(2!)} + \cdots \right] \tag{3.63}$$

For small values of β (large time values), the higher order terms in Eq. (3.63) become negligible. This can be achieved either by making the probe radius as small as possible or by making the test duration as long as possible (Eq. 3.60). The simplified equation for the line heat source method is obtained if the temperature versus time data are collected within a specific time interval $(t - t_0)$ (van der Held & van Drunen, 1949) as:

$$\Delta T - \Delta T_0 = \frac{Q}{4\pi k} \ln \left(\frac{t}{t_0} \right) \tag{3.64}$$

The rate of rise in temperature of the sample is a function of thermal conductivity of the material. A plot of $(\Delta T - \Delta T_0)$ versus $\ln(t/t_0)$ should yield a straight line and thermal conductivity can be found from the slope. The initial time (t_0) is arbitrarily chosen but it will simplify the data analysis if it is set equal to the time when the time–temperature plot in a semilogarithmic axis starts to become linear. Based on visual inspection, the initial data points are discarded to eliminate any transient initial effects.

The thermal conductivity can be expressed as:

$$k = \frac{Q}{4\pi} \frac{\ln(t/t_0)}{(\Delta T - \Delta T_0)} \tag{3.65}$$

The sources of error with the line heat source probe method can be summarized as follows:

1. One source of error is due to finite probe length since the line heat source theory is based on a line heat source of infinite length. Heat flow in axial direction must be considered if the probe length is not long enough as compared to its diameter. Murakami, Sweat, Sastry, Kolbe, Hayakawa, and Datta (1996b) have shown that error reduces to $\leq 0.1\%$ if the ratio of the probe length to diameter is greater than 30.

2. Finite probe diameter and the fact that the probe has different thermal properties than the samples may cause error. Van der Held and van Drunen (1949) introduced the time correction factor, t_c, which they subtracted from each time observation to correct for the effect of finite heater diameter and any resistance to heat transfer between the heat source and sample. Nix et al. (1967) reported that the time correction factor will be greater for larger probes because the deviation from the line heat source theory is greater for larger probes. The time correction factor is not necessary for proper probe design and also for food samples at temperatures above freezing because the specific heat of the probe is about the same as that of the food sample.

   To determine the time correction factor, t_c, first temperature data are plotted with respect to time using arithmetic scales. Then, the instantaneous slope dT/dt is taken at several different times from this plot. Next, dT/dt values against time are plotted on an arithmetic scale. The intercept on the time axis gives the time correction factor t_c, at which the rate of change of temperature dT/dt becomes zero. If a time correction factor is required, the thermal conductivity is expressed as:

$$k = \frac{Q \ln\left[(t - t_c)/(t_0 - t_c)\right]}{4\pi (T - T_0)} \tag{3.66}$$

   where

   k = effective thermal conductivity (W/m °C),
   t = time since the probe heater is energized (s),
   t_0 = initial time (s),
   t_c = time correction factor (s),
   T, T_0 = temperatures of probe thermocouple at time t and t_0, respectively (°C).

3. This method cannot be used for thin samples because the probe must be surrounded by a sufficient layer of sample (Sweat & Haugh, 1974). Finite sample size can cause errors if the sample boundaries experience a temperature change during measurement. The measurement time can be shortened and sample diameter can be increased to minimize the error arising from this situation. Excessive test duration also creates error as a result of convection and moisture transfer.

   If the plot of temperature versus the natural logarithm of time is linear ($r^2 > 0.99$), it means the sample diameter is sufficient. The ratio of the probe diameter to sample (container) diameter must be less than 1/30 to minimize the error (Drouzas & Saravacos, 1988). In addition, if the $4\alpha t/d^2 < 0.6$ (in which α is the thermal diffusivity, t is the duration of experiment, and d is the diameter of the cylinder), error is negligible according to Vos (1955).

4. Power input should be low. This is particularly important for frozen foods. Materials having higher thermal conductivity require higher power levels to obtain sufficient increase in temperature.

   High power inputs tend to cause localized fusion, resulting in highly variable properties over the experimental period. In addition, it creates a high temperature gradient, which may also cause moisture migration and heat convection. Sweat (1995) recommended that the power level should be selected depending on the temperature rise, which is necessary to get a high correlation between T and $\ln t$.

5. Truncation of infinite series of solution may also cause error. This error is minimized if the probe size is minimized and the experimental duration is optimized (the error is maximum at $t = t_0$ but decreases thereafter).

6. Contact resistance between sample and probe may cause error. This error shifts the time–temperature plot but does not change the slope. Therefore, no correction is required.

7. Convection effects may occur with liquid samples, but confining the analysis to the linear portion of the curve can eliminate it. It cannot be used with nonviscous liquids owing to convection currents which develop around the heated probe. Convection becomes important if PrGr > 100–1200 (van der Held & van Drunen, 1949) at which a deviation of straight line (T vs. $\ln t$) starts. Agar may be used in the samples to form a gel to minimize convection effects.

The probe method is not appropriate for determination of thermal conductivity at temperatures slightly below the initial freezing temperatures because of the large variation caused by ice formation. A longer probe and larger sample diameter may be required for measuring the thermal conductivity of ice using the line heat source probe method because of the high thermal diffusivity of ice.

Although it is simple and fast, the line heat source probe method performs local measurements and therefore relies on the assumption that the food is homogeneous. Therefore, this method is not suitable for porous foods.

Example 3.2. The line heat source probe method was used to determine the thermal conductivity of Red Delicious apples. The sample container was filled with the sample with the probe inserted at the center and it was placed in a constant temperature bath at 21°C for equilibration. After equilibrium was reached, the probe heater was activated. The electrical resistance of heated wire was 223.1 Ω and the electrical current through the heated wire was measured as 0.14 A. Calculate the thermal conductivity of Red Delicious apples from the time–temperature data recorded (Table E.3.2.1).

Table E.3.2.1. Time–Temperature Data Recorded Using Line Heat Source Probe Method for Red Delicious Apple

Time (s)	Temperature ($^\circ$C)
5	21.00
10	21.51
15	21.72
20	21.97
25	22.15
30	22.29
35	22.33
40	22.44
45	22.57
50	22.63
55	22.72
60	22.77

Solution:

The equation for the line heat source method is:

$$\Delta T - \Delta T_0 = \frac{Q}{4\pi k} \ln \left(\frac{t}{t_0}\right) \tag{3.64}$$

From the linear regression of $(\Delta T - \Delta T_0)$ versus $\ln (t/t_0)$, the slope of $Q/4\pi k$ is obtained as 0.718 K ($r^2 = 0.995$).

The heat supplied per unit length (Q) is calculated from electrical resistance of heated source and the electrical current:

$$Q = I^2 R$$
$$Q = (0.14)^2 (223.1) = 4.37 \, \text{W/m}$$

Then, from the slope, thermal conductivity of Red Delicious apples is calculated as:

$$\text{Slope} = \frac{Q}{4\pi k} = 0.718 \, \text{K}$$
$$\Rightarrow k = 0.499 \, \text{W/m} \cdot \text{K}$$

(b)   Transient Hot Wire Method. This method involves a thin heater wire similar to the line heat source method. However, in this case, the hot wire is located at the interface between the sample and a reference of known thermal conductivity (Fig. 3.8). The heater and temperature sensor in the hot wire thermal conductivity apparatus consist of a single wire that is exposed to the material, while in the thermal conductivity probe there are separate wires that are usually sealed in a tube.

When electrical power is applied to the heater wire, the temperature rise ΔT at a point located on the interface between the two materials at a distance x from the heater wire may be described as (Takegoshi, Imura, Hirasawa, & Takenaka, 1982):

$$\Delta T = -\frac{Q\kappa\sigma}{4\pi k_1} \int_0^1 \frac{E_1 \left\{ -\kappa^2 x^2 \big/ 4\alpha_1 t \left(\kappa^2 u + 1 + u\right) \right\}}{\left(\kappa^2 u + 1 - u\right)^{1/2} \left(1 - u + \sigma^2 u\right)^{3/2}} du \tag{3.67}$$

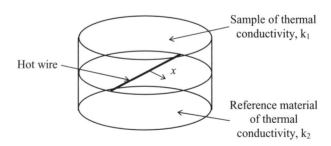

Figure 3.8   Transient hot wire method. [From Sastry, S.K., & Cornelius, B.D. (2002). *Aseptic Processing of Foods Containing Solid Particulates.* New York: John Wiley & Sons. Copyright © (2002) with permission from John Wiley.]

where $E_1\{\}$ is an integral exponential function and:

$$\kappa = \sqrt{\left(\frac{\alpha_1}{\alpha_2}\right)} \tag{3.68}$$

$$\sigma = \frac{k_2\sqrt{\alpha_1}}{k_1\sqrt{\alpha_2}} \tag{3.69}$$

where k is the thermal conductivity and α is the thermal diffusivity. The subscripts 1 and 2 refer to the thermal properties of sample and reference, respectively.

If x is small, Eq. (3.67) may be simplified to (Sharity-Nissar, Hozawa, & Tsukuda, 2000):

$$\frac{d\Delta T}{d\ln t} = \frac{Q}{2\pi\,(k_1 + k_2)}\left\{1 - \frac{\left(k_1\dfrac{\alpha_2}{\alpha_1} + k_2\right)}{k_1 + k_2}\frac{x^2}{4\alpha_2 t}\right\} \tag{3.70}$$

If the thermal diffusivities of sample and reference are the same ($\alpha_1 \approx \alpha_2$), and $\dfrac{x^2}{4\alpha_2 t} < 10^{-2}$, the thermal conductivity of the sample can be determined from:

$$k_1 = \frac{Q}{2\pi}\frac{d\ln t}{dT} - k_2 \tag{3.71}$$

Note that if the two materials are the same (i.e., $k_1 = k_2$), the equation reduces to the thermal conductivity probe equation (Eq. 3.65).

Sharity-Nissar et al. (2000) used a ribbon hot wire of width h and an infinitesimal thickness instead of a line heat source. The solution is then:

$$\frac{d\Delta T}{d\ln t} = \frac{Q}{2\pi\,(k_1 + k_2)}\left\{1 - \left(1 + \frac{h^2}{12x^2}\right)\frac{\left(k_1\dfrac{\alpha_2}{\alpha_1} + k_2\right)}{k_1 + k_2}\frac{x^2}{4\alpha_2 t}\right\} \tag{3.72}$$

Sharity-Nissar et al. (2000) indicated that the method is more convenient for measurements under high pressures than the thermal conductivity probe method. One obvious advantage is that probe insertion is no longer an issue, because the heater wire is sandwiched between two different materials. The method works best if the thermal diffusivities of the two materials are nearly equal. Thus, it may be necessary to change the reference for each sample. Because of the similarity of this method to the line heat source method, many of the error sources would be expected to be similar.

(c)   Modified Fitch Method. One of the most common transient methods used to measure the thermal conductivity of low conductivity materials is the Fitch method (Fitch, 1935). The Fitch method consists of a heat source or a sink in the form of a vessel filled with constant temperature liquid, and a sink or a source in the form of a copper plug insulated on all sides except one face through which heat transfer occurs. The sample is sandwiched between the vessel and the open face of the plug. Then, the temperature of the plug varies with time depending on the heat flow rate through the sample. Copper may be considered as lumped system since its thermal conductivity is high enough, and its temperature history may be used together with its mass and physical properties for calculation of the sample thermal conductivity (Zuritz, Sastry, Mccoy, Murakami, & Blaisdell, 1989).

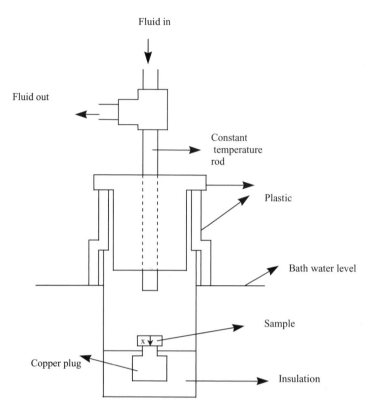

Figure 3.9 Cross-section of the modified Fitch apparatus. [From Sahin, S., Sastry, S.K., & Bayındırlı, L. Effective thermal conductivity of potato during frying: Measurement and modelling. *International Journal of Food Properties, 2,* 151–161. Copyright © (1999) with permission from Taylor & Francis.]

Bennett, Chace, and Cubbedge (1962) suggested a modified version of the commercially available Fitch apparatus for measuring thermal conductivity of soft materials such as fruits and vegetables. In the modification, a screw mechanism was provided to control the pressure on the sample.

Zuritz et al. (1989) modified the Fitch apparatus to make the device suitable for measuring thermal conductivity of small food particles that can be formed into slabs. The cross section of the modified Fitch apparatus is shown in Fig. 3.9.

In the Fitch apparatus, the sample is placed on the copper plug and equilibrated to room temperature. The initial sample temperature is recorded prior to the test. Then, the copper rod, the temperature of which is kept constant at a higher value than that of the sample and copper plug by circulating a fluid, is lowered and contacted with the sample. Time versus temperature data are recorded at the same time.

One-dimensional heat transfer occurs through the slab. Assuming constant properties, the governing differential equation for the temperature field within the sample is expressed as:

$$\frac{\partial T}{\partial t} = \alpha \frac{\partial^2 T}{\partial x^2} \tag{3.73}$$

where α is the thermal diffusivity of the sample.

The sample is considered to be at an initial temperature of T_0. One face of the sample is maintained at constant temperature (in contact with the vessel) and the other in contact with a perfect conductor (copper plug). Therefore, the differential equation (Eq. 3.74) can be solved with the following initial and boundary conditions:

I.C. at $t = 0$, $T = T_0$ for $0 < x < L$ (3.74)

B.C.1. at $x = 0$, $T = T_{\text{Cu-rod}}$ for $t > 0$ (3.75)

B.C.2. at $x = L$, $-k\dfrac{\partial T}{\partial x} = m_{\text{Cu-plug}} c_{p\text{Cu-plug}} \dfrac{\partial T}{\partial t}$ for $t > 0$ (3.76)

where

$c_{p\text{Cu-plug}}$ = specific heat of the cupper plug (J/kg K),
k = thermal conductivity of the sample (W/m K),
L = sample thickness (m),
$m_{\text{Cu-plug}}$ = mass of cupper plug (kg),
T_0 = initial temperature of both sample and copper plug (°C),
$T_{\text{Cu-rod}}$ = temperature of cupper rod (°C).

The analytical solution that satisfies the above differential equation is given by Carlslaw and Jaeger (1959) as:

$$T = T_{\text{Cu-rod}} + (T_0 - T_{\text{Cu-rod}}) \sum_{n=1}^{\infty} \frac{2\left(\lambda_n^2 + h^2\right) \exp\left(-\alpha\lambda_n^2 t\right) \sin(\lambda_n L)}{\lambda_n \left[L\left(\lambda_n^2 + h^2\right) + h\right]} \tag{3.77}$$

where

$$h = \frac{\rho c_p}{m_{\text{Cu-plug}} c_{p\text{Cu-plug}}} \tag{3.78}$$

and λ_n are the roots of:

$$\lambda_n \tan(\lambda_n L) = h \tag{3.79}$$

The assumption of quasi-steady conduction heat transfer through the sample, thereby ignoring the energy storage in the sample, yields the following simplified equation (Fitch, 1935; Mohsenin, 1980):

$$\frac{kA(T - T_{\text{Cu-rod}})}{L} = m_{\text{Cu-plug}} c_{p\text{Cu-plug}} \frac{dT}{dt} \tag{3.80}$$

Integrating Eq. (3.80) from $t = 0$ to any time t:

$$\int_0^t \frac{kA}{L\, m_{\text{Cu-plug}}\, c_{p\text{Cu-plug}}} dt = \int_{T_0}^{T} \frac{dT}{T - T_{\text{Cu-rod}}} \tag{3.81}$$

$$\Rightarrow \ln\left(\frac{T_0 - T_{\text{Cu-rod}}}{T - T_{\text{Cu-rod}}}\right) = \frac{kAt}{Lm_{\text{Cu-plug}} c_{p\text{Cu-plug}}} \tag{3.82}$$

where

A = heat transfer area (m²),
T = temperature of both copper plug and sample at time t (°C),
$T_{\text{Cu-rod}}$ = temperature of copper rod (°C).

As can be seen in Eq. (3.82), a plot of $\ln\left(\frac{T_0 - T_{\text{Cu-rod}}}{T - T_{\text{Cu-rod}}}\right)$ versus time is a straight line and thermal conductivity is calculated from the slope. Data must be analyzed in the linear temperature history region. Based on visual inspection, the initial data points are discarded to eliminate transient initial effects. A satisfactory fit was arbitrarily defined as a straight line with coefficient of determination $(r^2) \geq 0.995$. When r^2 is below this value, the topmost points were discarded, one point at a time until the r^2 is at least 0.995.

Mohsenin (1980) suggested that each Fitch apparatus should be calibrated with standard equipment such as a heat flow meter. The correction factor (β) for the apparatus can be calculated from the ratio between the thermal conductivity determined using standard apparatus and the thermal conductivity measured using the Fitch method. Then, the thermal conductivity of any material using the Fitch method is equal to the measured value multiplied by the correction factor.

While the analytical solution is the more accurate and general one, the simplified model is more frequently chosen because of its simplicity and ease of use.

Equation (3.80) is valid only if the following assumptions hold:

1. Contact resistance is negligible: This is the most important source of error in this method and it is especially important for high thermal conductivity materials. Good contact can be achieved if the contact surfaces are smooth and the sample faces are parallel. For rigid samples, pressure can be applied to provide good contact. However, for soft samples, it may not be possible. For these kinds of samples, application of a thin layer of nonwetting liquid to eliminate the air gaps between contact surface and porous materials is suggested.

2. Heat storage in the sample is negligible (i.e., heat transfer in the sample is quasi-steady state): For a given material, the sample thickness should be as small as possible to make a quasi-steady-state assumption. For the heat storage in the sample to be negligible with respect to that in the copper plug, the following relationship must hold:

$$\frac{m_s c_{p_s}\left(dT_s/dt\right)}{m_{\text{Cu-plug}} c_{p_{\text{Cu-plug}}}\left(dT_c/dt\right)} \ll 1 \tag{3.83}$$

where

m_s, $m_{\text{Cu-plug}}$ = mass of sample and copper plug, respectively (kg),

c_{p_s}, $c_{p_{\text{Cu-plug}}}$ = specific heat of sample and copper plug, respectively (J/kg K).

Since the time rates of temperature change for both sample and the copper plug are of the same order of magnitude, the following relationship is obtained:

$$\frac{m_s c_{p_s}}{m_{\text{Cu-plug}} c_{p_{\text{Cu-plug}}}} \ll 1 \tag{3.84}$$

The mass of the sample can be written as:

$$m_s = \rho_s\left(\pi R^2 L_{\max}\right) \tag{3.85}$$

where

$L_{\max}$ = the upper limit of optimum sample thickness (m),

R = radius of sample (m),

ρ_s = density of sample (kg/m^3).

The upper limit of the optimum sample thickness, L_{max}, can be estimated by substituting Eq. (3.85) into Eq. (3.84).

3. The heat transfer at the edges of the sample and copper plug is negligible: The error coming from this assumption is minimal if the temperature gradient is small, the sample is thin (area of heat transfer to the air is small), and thermal conductivity of the surroundings is low.

4. The temperature of the copper rod is constant: This can be achieved by contact with a large thermal mass at constant temperature and by using a vacuum flask.

5. The initial temperature of the copper plug and the sample are the same: This can be achieved by holding the sample and the copper plug at the same temperature environment prior to the experiment.

6. The temperature distribution in the copper plug is uniform: The high thermal conductivity and the small size of plug satisfy this condition. For the lumped system assumption to be valid, the internal to external heat transfer resistances of copper plug should be less than 0.1.

7. The sample is homogeneous across the heat transfer area: A larger apparatus size may be necessary for porous materials with pore sizes comparable to the heat transfer area.

Example 3.3. Thermal conductivity of an apple sample is measured by using a Fitch apparatus consisting of a copper rod and a copper plug. The sample was shaped to obtain a disk with 7.5 mm diameter and 3.0 mm thickness. The initial temperature of both sample and copper plug was 25°C. After equilibrium was reached, the copper rod, which has a constant temperature at 35°C, was lowered, making good contact with the sample surface and the temperature variation of copper plug was recorded (Table E.3.3.1).

(a) Calculate the thermal conductivity of the apple sample if the mass and specific heat of copper plug are 12.0 g and 385 J/kg.°C, respectively.

(b) Is it reasonable to assume that the heat storage in the sample is negligible? (The mass and specific heat values of the sample are 0.32 g and 4019 J/kg°C, respectively and assume that the time rate of temperature change for both the sample and the copper plug are the same.)

Solution:

(a) The equation for the modified Fitch method is:

$$\ln\left(\frac{T_0 - T_{Cu-rod}}{T - T_{Cu-rod}}\right) = \frac{kAt}{Lm_{Cu-plug}c_{pCu-plug}} \tag{3.82}$$

Table E.3.3.1. Time–Temperature Data Recorded Using Fitch Method for Apple

Time (s)	Temperature (°C)
0	25.00
5	25.08
10	25.16
15	25.24
20	25.32
25	25.39
30	25.47

Table E.3.3.2. $\ln\left(\dfrac{T_0 - T_{Cu-rod}}{T - T_{Cu-rod}}\right)$ Values as a Function of Time.

Time (s)	$\ln\left(\dfrac{T_0 - T_{Cu-rod}}{T - T_{Cu-rod}}\right)$
0	0
5	0.008032
10	0.016129
15	0.024293
20	0.032523
25	0.039781
30	0.048140

Heat transfer area is:

$$A = \pi \left(\frac{7.5 \times 10^{-3} m}{4}\right)^2 = 4.4 \times 10^{-5}\, \text{m}^2$$

Using the data, $\ln\left(\dfrac{T_0 - T_{Cu-rod}}{T - T_{Cu-rod}}\right)$ is calculated as a function of time (Table E.3.3.2).

Plotting $\ln\left(\dfrac{T_0 - T_{Cu-rod}}{T - T_{Cu-rod}}\right)$ versus time gives a straight line and thermal conductivity is calculated from the slope (Fig. E.3.3.1):

$$\text{Slope} = \frac{kA}{Lm_c C_{pc}} = 0.0016\,\text{s}^{-1}.$$

$$\Rightarrow k = \frac{(0.0016)(3 \times 10^{-3})(12 \times 10^{-3})(385)}{4.4 \times 10^{-5}}$$

$$k = 0.504\,\text{W/m}^\circ\text{C}$$

(b) If the heat storage in the sample is negligible:

$$\frac{m_s c_{p_s}(dT_s/dt)}{m_{Cu-plug} c_{p_{Cu-plug}}(dT_c/dt)} \ll 1 \tag{3.83}$$

$$\frac{dT_s}{dt} \cong \frac{dT_c}{dt}$$

$$\frac{m_s c_{p_s}}{m_{Cu-plug} c_{p_{Cu-plug}}} \ll 1 \tag{3.84}$$

$$\frac{(0.32)(4019)}{(12)(385)} = 0.28 \ll 1$$

Therefore, it is reasonable to assume that heat storage in the apple sample is negligible.

(d) Point Heat Source Method. This method involves a point heat source, which is heated for a period of time followed by monitoring of its temperature as the heat dissipates through the sample. The typical device used for this purpose is a thermistor that serves as both a heating element and a

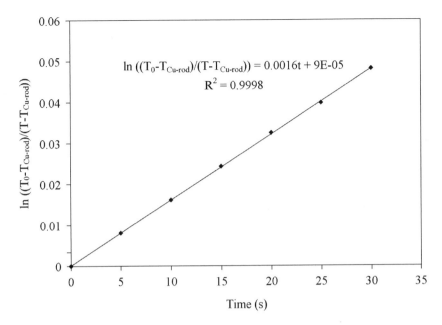

Figure E.3 3.1 Plot of $\ln\left(\dfrac{T_0 - T_{Cu-rod}}{T - T_{Cu-rod}}\right)$ versus time for calculation of thermal conductivity.

temperature sensor. Voudouris and Hayakawa (1994) conducted a detailed theoretical analysis of this method to determine the lower limits on sample size.

The analytical solution is given in the following equation (Carslaw & Jaeger, 1959; Voudouris & Hayakawa, 1994):

$$T - T_0 = \frac{Q}{4\pi kr} erfc\left(\frac{r}{\sqrt{4\alpha t}}\right) \tag{3.86}$$

where

$$erfc(x) = 1 - erf(x) = 1 - \frac{2}{\sqrt{\pi}} \int_0^x \exp(-\eta^2)d\eta = 1 - \frac{2}{\sqrt{\pi}} \int \left(1 - \frac{2}{2!}\eta^2 + \frac{12}{4!}\eta^4 - \cdots\cdots\right)$$

$$= 1 - \frac{2x}{\sqrt{\pi}} + \frac{2x^3}{\sqrt[3]{\sqrt{\pi}}} - \cdots \tag{3.87}$$

For small values of $r/\sqrt{4\alpha t}$, the higher order terms of complementary error function (erfc) drop out and Eq. (3.87) becomes:

$$T - T_0 = \left(\frac{Q}{4\pi k}\right)\left(\frac{1}{r} - \frac{1}{\sqrt{\pi\alpha t}}\right) \tag{3.88}$$

Thus, if $(T - T_0)$ is plotted against $1/\sqrt{t}$, the thermal conductivity can be found from the intercept first and then the thermal diffusivity can be determined from the slope. The effective value of r must be determined for the point heat source by conducting experiments using reference materials with known thermal properties. The theoretical analysis of Voudouris and Hayakawa (1994) indicated that the smaller the thermistor size, the smaller the dimensions of the sample. Much smaller thermistors are necessary for a measurement that is competitive with the line heat source method.

(e) Comparative Method. The comparative method is simple and a range of sample sizes can be handled via this method. Overall thermal conductivity measurement is possible rather than local measurement. Therefore, it is suitable for porous foods such as cakes (Carson, Lovatt, Tanner, & Cleland, 2004). It was successfully applied to the measurement of thermal conductivity of a variety of porous food analogues having a porosity between 0 and 0.65 and a mean pore size relative to the size of the sample container between 10^{-6} and 4×10^{-3} (volume basis) (Carson et al., 2004).

This method involves cooling of two spheres side by side in a well stirred ice/water bath. One sphere contains the sample and the other contains a reference of known thermal conductivity. The thermal conductivity of the sample is calculated from the time–temperature data of the cooling spheres. It is based on the analytical solution for the center temperature of a sphere being cooled with convection boundary conditions as described by:

$$\frac{T - T_\infty}{T_0 - T_\infty} = \sum_{i=1}^{\infty} \frac{2\text{Bi}\left(\beta_i^2 + (\text{Bi} - 1)^2\right)\sin\beta_i}{\beta_i\left(\beta_i^2 + \text{Bi}(\text{Bi} - 1)\right)} \exp\left(-\beta_i^2 \text{Fo}\right) \tag{3.89}$$

where
   Fo and Bi are the Fourier and Biot numbers, respectively, and
   T_0 = initial temperature (°C),
   T_∞ = bulk fluid temperature (°C).
   The β_i coefficients are the roots of $\beta_i \cot \beta_i + (Bi - 1) = 0$.
   The solution involves infinite series which are difficult to deal with. However, the terms in the solution converge rapidly with increasing time. Approximating the infinite series in Eq. (3.89) by the first term alone for Fo > 0.2 and using the temperature–time data, the thermal conductivity of the sample can be determined.

The temperature–time histories of the cooling spheres are plotted on a logarithmic scale and the slope of the linear portion of the graph is determined:

$$\ln\left(\frac{T - T_\infty}{T_0 - T_\infty}\right) = B - \frac{\beta_1^2 \alpha}{R^2} t \tag{3.90}$$

Calculations are made for the sample and reference spheres relative to each other. Then:

$$\frac{k_{\text{sample}}}{k_{\text{ref}}} = \frac{\text{Slope}_{\text{sample}}}{\text{Slope}_{\text{ref}}} \frac{\rho}{\rho} \frac{c_{p_{\text{sample}}}}{c_{p_{\text{ref}}}} \frac{R_{\text{sample}}^2}{R_{\text{ref}}^2} \frac{\beta_{1_{\text{ref}}}^2}{\beta_{1_{\text{sample}}}^2} \tag{3.91}$$

The end losses are eliminated by using spherical sample containers in measurements. However, the measurement takes a long time. The temperature dependence of the thermal conductivities of the materials involved is not determined and instead temperature averaged conductivity values are measured. It is an indirect method for thermal conductivity measurement. Density and specific heat data are required since the method directly gives thermal diffusivity.

3.3 SPECIFIC HEAT

Specific heat is the amount of heat required to increase the temperature of a unit mass of the substance by unit degree. Therefore, its unit is J/kg K in the SI system. The specific heat depends on the nature of the process of heat addition in terms of either a constant pressure process or a constant volume process. However, because specific heats of solids and liquids do not depend on pressure much, except extremely high pressures, and because pressure changes in heat transfer problems of agricultural materials are usually small, the specific heat at constant pressure is considered (Mohsenin, 1980).

The specific heats of foodstuffs depend very much on their composition. Knowing the specific heat of each component of a mixture is usually sufficient to predict the specific heat of the mixture (Sweat, 1995).

Specific heat data for different food materials below and above freezing were given in Rahman (1995) and Singh (1995).

3.3.1 Prediction of Specific Heat

The specific heat of high-moisture foods is largely dominated by water content. As early as 1892, Siebel argued that the specific heat of the food can never be greater than the sum of the specific heat of the solid matter and water since water in food materials exists side by side with the solid matter without any heat producing chemical reactions (Siebel, 1892). In addition, the specific heat of food materials cannot be much smaller than that of water since water is the major component of food materials. Siebel (1892) proposed the following equation for aqueous solutions such as vegetable and fruit juices or pastes:

$$c_p = 0.837 + 3.349 X_w^w \tag{3.92}$$

Siebel (1892) also suggested the following equation for food materials below their freezing point:

$$c_p = 0.837 + 1.256 X_w^w \tag{3.93}$$

where X_w^w is the mass fraction of moisture within the sample and specific heat, and c_p is given in kJ/kg K.

The reason for the lower values of c_p below freezing is that the specific heat of ice is about one half of that of the liquid water. This also partly explains the higher thawing times of foods as compared to their freezing times.

Heldman (1975) proposed the following equation to estimate the specific heat of foodstuffs using the mass fraction of its constituents (water, protein, fat, carbohydrate, and ash):

$$c_p = 4.180 X_{water}^w + 1.547 X_{prot}^w + 1.672 X_{fat}^w + 1.42 X_{CHO}^w + 0.836 X_{ash}^w \tag{3.94}$$

Choi and Okos (1986) suggested the following equation for products containing n components:

$$c_p = \sum_{i=1}^{n} X_i^w c_{pi} \tag{3.95}$$

where

X_i^w = mass fraction of component i,
c_{p_i} = specific heat of component i (J/kg K).

The temperature dependence of specific heat of major food components has been studied. The specific heat of pure water, carbohydrate (CHO), protein, fat, ash, and ice at different temperatures can be expressed empirically in J/kg°C according to Choi and Okos (1986) as follows:

$$c_{p_{water}} = 4081.7 - 5.3062\,T + 0.99516T^2 \qquad \text{(for } -40\text{ to }0°C) \tag{3.96}$$

$$c_{p_{water}} = 4176.2 - 0.0909\,T + 5.4731 \times 10^{-3}T^2 \qquad \text{(for 0 to }150°C) \tag{3.97}$$

$$c_{p_{CHO}} = 1548.8 + 1.9625\,T - 5.9399 \times 10^{-3}T^2 \qquad \text{(for } -40\text{ to }150°C) \tag{3.98}$$

$$c_{p_{protein}} = 2008.2 + 1.2089\,T - 1.3129 \times 10^{-3}T^2 \qquad \text{(for } -40\text{ to }150°C) \tag{3.99}$$

$$c_{p_{fat}} = 1984.2 + 1.4373\,T - 4.8008 \times 10^{-3}T_2 \qquad \text{(for } -40\text{ to }150°C) \tag{3.100}$$

$$c_{p_{ash}} = 1092.6 + 1.8896\,T - 3.6817 \times 10^{-3}T^2 \qquad \text{(for } -40\text{ to }150°C) \tag{3.101}$$

$$c_{p_{ice}} = 2062.3 + 6.0769\,T \tag{3.102}$$

where temperature (T) is in (°C) in these equations.

Specific heat of moist air can be expressed as a function of relative humidity (RH) of air (Riegel, 1992):

$$c_{p_{moist\,mair}} = c_{p_{dry\,air}}(1 + 0.837RH) \tag{3.103}$$

Generally, experimentally determined specific heat is higher than the predicted value. The reason may be the presence of bound water, variation of specific heat of the component phases with the source and interaction of the component phases (Rahman, 1995). Rahman (1993) considered the excess specific heat, c_{ex}, due to the interaction of the component phases and proposed the following equation:

$$c_p = \left[\sum_{i=1}^{n} X_i^w c_{pi}\right] - c_{ex} \tag{3.104}$$

Rahman (1993) correlated the excess specific heat for fresh seafood as:

$$c_{ex} = -33.77 + 85.58\left(X_w^w\right) - 53.76\left(X_w^w\right)^2 \tag{3.105}$$

These equations are valid in a temperature change when there is no phase change. If there is a phase change, latent heat must be incorporated. This is accomplished by using a new term, apparent specific heat, which includes both latent and sensible heat:

$$c_{p_{app}} = \frac{dH}{dT} \tag{3.106}$$

Example 3.4. Estimate the specific heat of potatoes containing 85% water.

Data:

$$c_{p_{water}} = \quad 4186.80\,\text{J/kg K}$$

$$c_{p_{nonfat\,solids}} = 837.36\,\text{J/kg K}$$

Table E.3.5.1. Approximate Composition of Rice

Component	Weight (%)
Water	8.5
Carbohydrate	75.3
Protein	14.1
Fat	0.7
Ash	1.4

Solution:

Using Eq. (3.95) as suggested by Choi and Okos (1986):

$$c_p = \sum_{i=1}^{n} X_i^w c_{pi} \qquad (3.95)$$

$$c_p = (0.85)(4186.8) + (0.15)(837.36)$$

$$= 3684.38 \text{ J/kg K}$$

Example 3.5. Calculate the specific heat of wild rice grain at 20°C with the approximate composition data given in Table E.3.5.1.

Solution:

Using the specific heat correlations (Choi & Okos, 1986), specific heat of each component at 20°C can be calculated. The results are tabulated in Table E.3.5.2.

Introducing these values into Eq. (3.95) and using the composition of rice, c_p of rice grain is determined.

$$c_p = \sum_{i=1}^{n} X_i^w c_{pi} \qquad (3.95)$$

$$c_p = (0.085)(4176.6) + (0.753)(1585.7) + (0.141)(2031.9) + (0.007)(2011) + (0.014)(1128.9)$$

$$= 1865.4 \text{ J/kg}°C$$

Table E.3.5.2. Predicted Specific Heat Values of Components at 20°C

Component	Specific Heat Equation	Eq. no.	c_{p_i} (J/kg°C)
Water	$c_{p_{water}} = 4176.2 - 0.0909\,T + 5.4731 \times 10^{-3}\,T^2$	(3.97)	4176.6
Carbohydrate	$c_{p_{CHO}} = 1548.8 + 1.9625\,T - 5.9399 \times 10^{-3}\,T^2$	(3.98)	1585.7
Protein	$c_{p_{protein}} = 2008.2 + 1.2089\,T - 1.3129 \times 10^{-3}\,T^2$	(3.99)	2031.9
Fat	$c_{p_{fat}} = 1984.2 + 1.4373\,T - 4.8008 \times 10^{-3}\,T^2$	(3.100)	2011.0
Ash	$c_{p_{ash}} = 1092.6 + 1.8896\,T - 3.6817 \times 10^{-3}\,T^2$	(3.101)	1128.9

3.3.2   Measurement of Specific Heat

Methods of mixture, guarded plate, comparison calorimeter, adiabatic agricultural calorimeter, differential scanning calorimeter (DSC), and calculated specific heat are some of the methods used for the determination of specific heat (Mohsenin, 1980).

3.3.2.1   Method of Mixture

The method of mixture is the most widely used system for measuring specific heat of food and agricultural materials because of its simplicity and accuracy. A known quantity of liquid (typically water) at a known initial temperature is mixed with a sample of known mass and temperature within an insulated container. The equilibrium temperature of the mixture is determined and the specific heat may be calculated from the simple energy balance (Mohsenin 1980):

$$m_c c_{p_c} (T_i - T_e) + m_w c_{p_w} (T_i - T_e) = m_s c_{p_s} (T_e - T_{is}) \tag{3.107}$$

where

m_c = mass of calorimeter cup (kg),
c_{p_c} = specific heat of calorimeter cup (J/kg K),
m_s = mass of sample (kg),
c_{p_s} = specific heat of sample (J/kg K),
m_w = mass of water (kg),
c_{p_w} = specific heat of water (J/kg K),
T_i = initial temperature of calorimeter and water (K),
T_{is} = initial temperature of sample (K),
T_e = equilibrium temperature of mixture (K),
T_w = initial temperature of water (K).

The accuracy of this method is based on the assumption of negligible heat exchange between the calorimeter and its surrounding atmosphere. In this method, an average c_p value is obtained from the corresponding temperature range. If a complete specific heat–temperature relationship is desired, additional experiments must be run at different temperature ranges.

To determine the specific heat of hygroscopic food materials, the sample is encapsulated in a copper cylinder. The density of the heat transfer medium should be lower than the food sample so that it submerges.

Example 3.6. A vacuum jacketed calorimeter of 86 g was used for the determination of specific heat of 36 g of lean beef sample. The calorimeter cup and the sample were first brought to 12°C and then 68 g of water at 6°C was poured into the cup. The system was then sealed and brought to equilibrium. The equilibrium temperature was recorded to be 7.8°C. The specific heats of water and calorimeter cup are 4198 J/kg°C and 383 J/kg°C, respectively. The calorimeter was well insulated and the heat loss to the surrounding was negligible. Calculate the specific heat of your sample.

Solution:

Assuming that there is no heat loss or gain to or from the surrounding:

$$\left(\begin{array}{c} \text{Amount of} \\ \text{Energy given} \end{array} \right)_{\text{calorimeter}} + \left(\begin{array}{c} \text{Amount of} \\ \text{Energy given} \end{array} \right)_{\text{sample}} = \left(\begin{array}{c} \text{Amount of} \\ \text{Energy received} \end{array} \right)_{\text{water}}$$

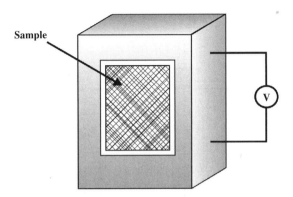

Figure 3.10   Guarded plate method.

$$(86 \times 10^{-3} \, \text{kg}) \left(383 \frac{J}{\text{kg} \cdot {}^\circ C}\right) (12 - 7.8)({}^\circ C) + (36 \times 10^{-3} \, \text{kg}) c_{p_s} (12 - 7.8)({}^\circ C)$$

$$= (68 \times 10^{-3} \, \text{kg}) \left(4198 \frac{J}{\text{kg} \cdot {}^\circ C}\right) (7.8 - 6)({}^\circ C)$$

$$c_{p_s} = 2483.4 \, J/\text{kg}{}^\circ C$$

3.3.2.2   Method of Guarded Plate

In this method, the sample is surrounded by electrically heated thermal guards that are maintained at the same temperature as the sample (Fig. 3.10). The sample is also being heated electrically; thus ideally there is no heat loss. The electric heat supplied to the sample in a given time (t) is set equal to the heat gain by the sample as:

$$mc_p(T_{es} - T_{is}) = VIt \tag{3.108}$$

where

V = voltage (V),
I = current (A),
m = mass of the sample (kg),
T_{is}, T_{es} = initial and final temperatures of the sample (${}^\circ C$).

3.3.2.3   Method of Comparison Calorimeter

The comparison calorimeter is used to determine the specific heat of liquids. There are two cups in the calorimeter (Fig. 3.11). One cup (cup A) is filled with distilled water or other liquids with known specific heat while the other cup (cup B) is filled with the sample liquid. Both cups are heated to the same temperature and then placed in the calorimeter to cool down. The temperature data for both liquids are taken at regular intervals as the liquids cool down. Cooling curves are obtained for both liquids and the rates of cooling are determined at the same temperature (Fig. 3.12). By comparing the cooling curves of the two liquids, specific heat of the sample is calculated. At a given temperature, the

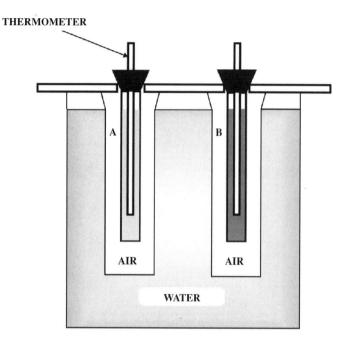

Figure 3.11 Comparison calorimeter.

rates of heat loss from the two liquids contained in calorimeter are equal. If the temperature change of the cooling body is sufficiently small, it can be assumed that the specific heats are constant and the rate of heat loss is equal to the rate of temperature change (Mohsenin, 1980). Thus:

$$\left(m_A c_{p_A} + m_w c_{p_w}\right) \frac{\Delta T}{\Delta t_A} = \left(m_B c_{p_B} + m_s c_{p_s}\right) \frac{\Delta T}{\Delta t_B} \tag{3.109}$$

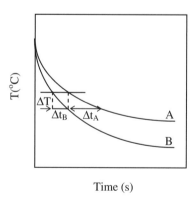

Time (s)

Figure 3.12 Cooling curves in comparison calorimeter.

Then, specific heat of the sample is:

$$c_{p_s} = \frac{\left(m_A c_{p_A} + m_w c_{p_w}\right) \Delta t_B - \left(m_B c_{p_B}\right) \Delta t_A}{m_s \Delta t_A} \tag{3.110}$$

where

ΔT = temperature drop for cups A and B (K),
m = mass (kg),
Δt = time for cup and its contents to drop ΔT (s),

The subscripts A, B, w, s denote cup A, cup B, water, and the sample, respectively.

3.3.2.4 Adiabatic Agricultural Calorimeter

The adiabatic calorimeter is designed in a way that there is no heat or moisture transferred through the sample chamber walls (Mohsenin, 1980). A measured quantity of heat is added by means of heating cables buried in the bulk of the sample in a container placed in a sample chamber. The heat given raises the temperature of the sample, the sample container, and the chamber walls. Then, the specific heat is calculated using the energy balance.

The specific heat of the sample can be calculated from the energy balance (Mohsenin, 1980):

$$Q = (mc_p \Delta T)_{\text{sample}} + (mc_p \Delta T)_{\text{container}} + (mc_p \Delta T)_{\text{chamber}} \tag{3.111}$$

There are two heating techniques in the adiabatic calorimeter: intermittent and continuous.

In the intermittent heating technique, a constant power input is applied to the sample heater for a carefully measured time interval (West & Ginnings, 1958). The adiabatic jacket is maintained at the same temperature as the sample during the heating period, providing adiabatic conditions between the sample and the environment. At the end of the heating period, the sample and the jacket temperatures are allowed to equilibrate and the final sample temperature is recorded. The sample's enthalpy–temperature relation is calculated from the difference between the initial and final temperatures. The thermocouples are located on inside and outside walls for monitoring temperatures.

The continuous heating technique involves running the sample heater and adiabatic jacket heaters continuously (Solomons & Gummings, 1964). The temperature and power inputs as a function of time are recorded. This technique provides a complete set of enthalpy–temperature results over a wide range of temperature in a relatively shorter time compared with the much longer and tedious intermittent heating technique. The major disadvantage of this method is the introduction of a thermal gradient within the sample which may become a source of significant error if enough care is not taken in the thermal design of the sample cell.

3.3.2.5 Differential Scanning Calorimeter (DSC)

Differential scanning calorimeter (DSC), which reports heat flow as a function of temperature is an excellent tool for the measurement of temperature-dependent specific heat and phase transitions. The measurement is made by heating a sample at a known and fixed rate. Once dynamic heating equilibrium of the sample is obtained, the heat flow is recorded as a function of temperature. This heat flow is directly proportional to the specific heat of the sample.

In this method, first thermal scan is conducted with an empty sample pan to obtain a baseline. Next, the sample is weighed and placed within a pan into the calorimeter receptacle and the same

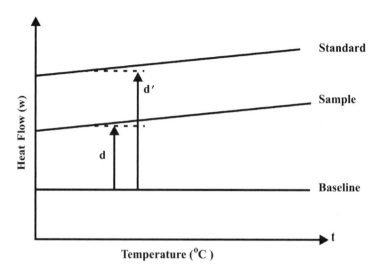

Figure 3.13   Thermogram for specific heat determination by DSC.

thermal scan is conducted again. A thermogram showing the rate of heat input versus temperature is obtained from the experimental data (Fig. 3.13). The deflection of the thermogram from the baseline is proportional to the differential rate of energy input (dH/dt) required to maintain the sample at the same temperature as the reference pan (Mohsenin, 1980):

$$d = \beta \frac{dH}{dt} \qquad (3.112)$$

where d is the deflection from the baseline with a sample in the pan and β is the proportionality constant.

Temperature is linearly related to time. Then:

$$d = \beta \frac{dH}{dT} \frac{dT}{dt} \qquad (3.113)$$

The specific heat of the sample (c_p) is given by:

$$c_p = \frac{1}{m} \frac{dH}{dT} \qquad (3.114)$$

Then, the deflection may be expressed as:

$$d = \beta m c_p \frac{dT}{dt} \qquad (3.115)$$

where dT/dt is the rate of temperature scanning.

Specific heat of sample can be determined if the proportionality constant β is known. If a reference material (usually sapphire) of known specific heat is used and the same thermal scan is conducted in DSC, the deflection from the baseline is:

$$d' = \beta m' c_p' \frac{dT}{dt} \qquad (3.116)$$

During the scanning period, any deflection from the base line would be directly proportional to the specific heat of the sample only. Therefore, the proportionality constant β is independent of the

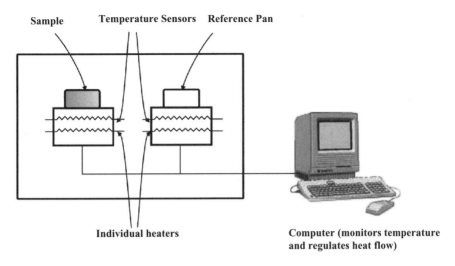

Sample   Temperature Sensors   Reference Pan

Individual heaters   Computer (monitors temperature and regulates heat flow)

Figure 3.14   Schematic representation of the DSC system.

specific heat values and will cancel out if the ratio of Eqs. (3.115) and (3.116) is taken. As a result:

$$c_p = \left(\frac{d}{d'}\right)\left(\frac{m'}{m}\right)c'_p \tag{3.117}$$

The DSC method offers several advantages. A small quantity of sample (10–20 mg) which minimizes the thermal lag within the system is used. It is quite suitable for calorimetric measurements since sample and reference are each provided in individual heaters, which allow the determination to be conducted with no temperature difference between the sample and the reference (Fig. 3.14). Thus, the calibration constant for the instrument is independent of temperature and the difference in heat flow to the sample and the reference in order to maintain equal temperature can be measured directly (Lund, 1984).

Calibration of the instrument is generally carried out with a high-purity metal having accurately known enthalpy of fusion and melting point. For calibration, mostly indium is used ($\Delta H_{fu} = 28.45$ kJ/kg: $T_{mp} = 156.4°C$) (Lund, 1984). The aluminum pans must be hermetically sealed to prevent evaporation of moisture from sample.

Example 3.7. Specific heat of granular starch having a moisture content of 23.08% is determined by using differential scanning calorimeter (DSC) over a temperature range of 0 to 180°C. The deflection from the baseline is read as 76.5 for the sample and 40 for the sapphire from the thermogram. The masses of the sample and sapphire are 25 and 82 mg, respectively. If the specific heat of sapphire is known as 0.194 kJ/kg°C, calculate the specific heat of granular starch.

Solution:

The specific heat of sample is calculated by using Eq. (3.117):

$$c_p = \left(\frac{d}{d'}\right)\left(\frac{m'}{m}\right)c'_p \tag{3.117}$$

$$= \left(\frac{76.5}{40}\right)\left(\frac{82}{25}\right)(0.194)$$

$$c_p = 1.217 \text{ kJ/kg°C}$$

3.3.2.6   Method of Calculated Specific Heat

The specific heat of food and agricultural materials can be determined from other thermal properties such as thermal conductivity and diffusivity. By using a constant-temperature heating method and a temperature distribution chart, a Fourier number can be obtained. From the Fourier number, the specific heat can be calculated in terms of thermal conductivity.

Gordon and Thorne (1990) developed a method for estimating thermal properties (thermal conductivity, specific heat, and heat transfer coefficient) of spherical foodstuffs from the density of the food and temperature–time data at two points within the food during cooling. Thermal properties were derived from the equations describing temperatures at two points within a sphere during cooling. The relationships between these variables and the derived quantities were evaluated by multiple linear regression and expressed as polynomials. The first roots of these expressions were extracted by the Newton-Rhapson technique and the expressions were then reduced in degree to yield subsequent roots. A routine was developed that selected the set of roots most appropriate to the original material. The specific heat values estimated by this method were similar to the published values.

3.4   ENTHALPY AND LATENT HEAT

Enthalpy is the heat content in a system per unit mass. Its unit is J/kg in SI system. It is a thermodynamic property that depends only on the state of the system and it is expressed in terms of internal energy, pressure, and volume as:

$$H = U + PV \tag{3.118}$$

Latent heat is the amount of heat released or absorbed at a specific temperature when unit mass of material transformed from one state to another.

DSC is the most common device for measuring state and phase transitions. When material undergoes a change in physical state, heat is liberated or absorbed. Phase transitions may result in peaks in the thermogram (Fig. 3.15). Peaks may be either endothermic such as in the case of melting, denaturation,

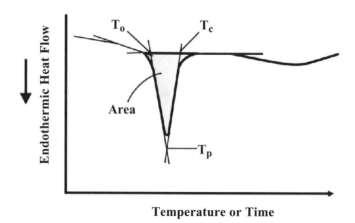

Figure 3.15   DSC endotherm for measuring area and temperatures (T_0, T_p, and T_C are the onset, peak and conclusion temperatures, respectively).

gelatinization, and evaporation or exothermic such as in freezing, crystallization, and oxidation. The area under the peak is directly proportional to the change of enthalpy (ΔH).

The thermal resistance to heat flow depends on the sample nature and packing and the extent of thermal contact (Lund, 1984). For optimum peak sharpness and resolution, the contact surface between pan and the sample should be maximized. This can be achieved by having samples as thin discs or films or fine granules. In the case of heterogeneous food materials, the sample must be homogenized.

Determination of phase transition enthalpy requires measurement of the endo- or exotherm area. If the baseline of the endo- or exotherm is not horizontal and peak is not symmetrical, it may be difficult to measure the area. As can be seen in Fig. 3.15, first the baseline on each side of the peak is extrapolated across the peak. The linear side of each side of the peak is extrapolated so that it will intersect the extrapolated baseline on its respective side. The intersections at points T_0 and T_C represent the initial and final temperatures of transitions. Peak area can be determined as depicted in the same figure.

It is possible to record gelatinization temperatures and enthalpy of gelatinization for different starch types using DSC. Different number of endotherms can be obtained depending on starch types by using DSC. When rice starch suspension having water to starch ratio greater than 2:1 is heated to 150°C at a rate of 10°C/min, two endotherms are obtained. The first endotherm represents phase transition of gelatinization while the second endotherm observed at higher temperatures (between 90 and 110°C) is associated with melting of amylose lipid complex (Lund, 1984). The peak at higher temperatures has not been reported in defatted starches or starches containing less than 0.1% lipids.

Starch retrogradation can be studied by DSC. In retrograded starch, the area under the peak represents the quantitative measure of energy transformation during the melting of recrystallized amylopectin as well as transition temperatures (T_0, T_p, and T_c). Starch retrogradation enthalpies (<8 J/g) were reported to be smaller than gelatinization enthalpies (9–15 J/g) (Abd Karim, Norziah, & Seow, 2000). Moreover, endothermic transitions associated with melting of starch takes places at temperatures 10 to 26°C lower than gelatinization temperature of starch.

3.5 THERMAL DIFFUSIVITY

Thermal diffusivity (α) is a physical property associated with transient heat flow. It is a derived property. The unit of thermal diffusivity is m^2/s in the SI system. It measures the ability of a material to conduct thermal energy relative to its ability to store thermal energy. Materials with large thermal diffusivity will respond quickly to changes in their thermal environment while materials of small thermal diffusivity will respond more slowly, taking longer to reach a new equilibrium condition. That is, the ratio of heating times of two materials of the same thickness will be inversely proportional to their respective diffusivities:

$$\frac{\alpha_1}{\alpha_2} = \frac{\Delta t_2}{\Delta t_1} \tag{3.119}$$

Then, thermal conductivity can be determined directly if the heating time through a given thickness of material is compared with the heating time of a material with known thermal diffusivity and the same thickness.

3.5.1 Indirect Prediction Method

Thermal diffusivity can be calculated indirectly from the measured thermal conductivity, density, and specific heat. This approach needs considerable time and different instrumentation.

3.5.2 Direct Measurement Methods

Direct measurement of thermal diffusivity is usually determined from the solution of one-dimensional unsteady-state heat transfer equation. The most commonly used thermal diffusivity measurement methods are the temperature history method, thermal conductivity probe method, and Dickerson method. The point heat source method discussed in the section on thermal conductivity measurement methods is also used to determine thermal diffusivity. Rahman (1995) reviewed several other measurement methods for thermal diffusivity.

3.5.2.1 The Temperature History Method

A heating or cooling experiment can be performed. Transient temperature history charts giving Fourier number as a function of temperature ratio for different Biot numbers are available and they are used to determine the thermal diffusivity (Heissler, 1947).

3.5.2.2 Thermal Conductivity Probe

A technique of adding two thermocouples at a known distance r laterally from the central thermocouple in a conductivity probe has been described for measurement of thermal diffusivity (McCurry, 1968; Nix et al., 1967). Ingersoll, Zobel, and Ingersoll (1954) gave the expression for cylindrical temperature field in the following form:

$$T = \frac{Q}{2\pi k} \int_{\beta}^{\infty} \frac{e^{-\beta^2}}{\beta} d\beta \tag{3.120}$$

where β is a dimensionless parameter defined as in the line heat source probe method:

$$\beta = \frac{r}{2\sqrt{\alpha t}} \tag{3.60}$$

The solution of Eq. (3.120) is (Nix et al., 1967):

$$T = \frac{Q}{2\pi k} \left[-\frac{C_e}{2} - \ln \beta + \frac{\beta^2}{2.1!} - \frac{\beta^4}{4.2!} + \cdots \right] \tag{3.121}$$

where C_e is the Euler constant, equal to 0.57721.

Equation (3.121) can be used to calculate thermal diffusivity by trial and error. To ensure convergence, the first 40 terms of equation have to be evaluated if the set of thermocouples were placed in such location that $0.16 < \beta < 3.1$ over the time interval used (Nix et al., 1967). This method is suitable for biological materials since the test duration is short (about 300 s) and the temperature change imposed on the sample is small. The exact distance r of the two thermocouples is critical. Therefore, a calibration of the system may be required.

3.5.2.3 Dickerson Method

Dickerson (1965) described an apparatus to measure thermal conductivity. The cylindrical container of radius R with high thermal conductivity is filled with sample and placed in a constant-temperature agitated water bath. Both ends of the container are insulated with rubber corks so that only a radial temperature gradient exists. The temperatures at the surface and center of the cylinder were monitored with thermocouples. The time–temperature data are recorded until constant rate of temperature rise is obtained for both the inner and outer thermocouples. Then, the Fourier equation is given as:

$$\frac{\Omega}{\alpha} = \frac{d^2T}{dr^2} + \frac{1}{r}\frac{dT}{dr} \tag{3.122}$$

where Ω is the constant rate of temperature rise at all points in the cylinder ($^{\circ}$C/s).
The solution of Eq. (3.122) is:

$$T = \frac{\Omega r^2}{4\alpha} + \omega \ln r + \varphi \tag{3.123}$$

B.C. 1 at $r = R$, the surface temperature is $T_s = T + \Omega t$ for $t > 0$ \hfill (3.124)

B.C.2 at $r = 0$ $\dfrac{dT}{dr} = 0$ for $t > 0$ \hfill (3.125)

Then, the solution of Eq. (3.123) is:

$$T_s - T = \frac{\Omega}{4\alpha}\left(R^2 - r^2\right) \tag{3.126}$$

at the center ($r = 0$), $T = T_c$. Then, thermal diffusivity can be expressed as:

$$\alpha = \frac{\Omega R^2}{4(T_s - T_c)} \tag{3.127}$$

In this method, the sample cylinder should have a length-to-diameter ratio of at least 4 to eliminate the axial heat transfer (Olson & Schultz, 1942).

PROBLEMS

3.1. The thermal conductivity of a Royal Gala apple is measured at 25°C by guarded hot plate method. The apple samples are cut into chips with area of 305 mm × 305 mm and thickness of 15 mm. The temperature difference between the hot and cold surfaces is kept at 2°C and the measured rate of heat input is 6.1 W. Calculate the thermal conductivity of the apple.

3.2. Determine the thermal conductivity of hazelnut having a porosity of 0.45 at 24.8°C using the isotropic model of Kopelman. Determine the thermal conductivity using the order: water (phase 1), carbohydrate (2), protein (3), fat (4), ash (5), and air (6). The composition data and density (ρ) and thermal conductivity (k) of each component are given in Table P.3.2.1.

3.3. The line heat source probe method was used to determine the thermal conductivity of a food sample. The sample container is filled with the sample with the probe inserted at the center and placed in a constant temperature bath at 25°C for equilibration. After the equilibrium is reached, the probe heater was activated. The electrical resistance of heated source per unit length is 223.1 Ω/m and the electrical current was measured as 0.14 A. Calculate the thermal conductivity from the data given in Table P.3.3.1.

3.4. A modified Fitch device was used for measuring the thermal conductivity of a strawberry sample. The sample was shaped to obtain a cylinder with 6.35 mm diameter and 2.25 mm height, which is suitable for the device. The initial temperature of both the sample and the copper plug was 25°C. After equilibrium was reached, the copper rod, which had a constant

Table P.3.2.1   Composition Data and Density and Thermal Conductivity of Each Component

Component	Composition (weight %)	ρ at 24.8 °C (kg/m³)	k at 24.8 °C (W/m K)
Water	5.6	995.0	0.62
Carbohydrate	12.3	1591.0	0.17
Protein	13	1317.0	0.21
Fat	66.5	915.0	0.23
Ash	2.6	2417.0	0.36
Air	Negligible weight	1.2	0.025

Table P.3.3.1   Temperature Versus Time Data During Measurement

Time (s)	Temperature (°C)
0	25.00
5	25.30
10	26.34
15	27.12
20	27.68
25	28.11
30	28.46

temperature of 35°C, was lowered, making good contact with the sample surface, and the temperature variation of copper plug was recorded and given in Table P.3.4.1.

Calculate the thermal conductivity of the sample if the mass and specific heat of the copper plug are 0.010762 kg and 384.9 J/kg K, respectively.

3.5. To determine the specific heat of the apple, 100 g of apple was heated from 20°C to 30°C for 10 min via the guarded plate method. The voltage and the current supplied to the heater were 0.5 V and 12 A, respectively. Calculate the specific heat of the apple.

3.6. The comparison calorimeter is used to determine the specific heat of milk. The first cup is filled with 100 g of distilled water having a specific heat of 4.18 kJ/kg K. The other cup is filled with 105 g of milk. Both cups are heated to the same temperature and then placed in the calorimeter to cool. The rates of cooling are 10°C/s and 10.3°C/s for water and milk, respectively. Both cups have the same mass (50 g) and specific heat of 0.95 kJ/kg K. The calorimeter is well insulated and the heat loss to the surroundings is negligible. Calculate the specific heat of milk.

3.7. The specific heat of cucumber was determined using DSC. An 18-mg sample was used. The deflection from the baseline was 87.78 for the sample and 19 for the sapphire at 32°C in the thermogram. The specific heat of the sapphire is 0.191 kJ/kg K. If the mass of reference material is 82 mg, calculate the specific heat of cucumber at 32°C.

3.8. The Dickerson method was used to measure thermal diffusivity of a sample. The variation of surface (T_s) and center (T_c) temperatures of a sample during measurement is shown in Table

Table P.3.4.1 Temperature Variation of Copper Plug During Measurement

Time (s)	Temperature (°C)
0	25.00
2	25.03
4	25.06
6	25.10
8	25.13
10	25.16

Table P.3.8.1 Variation of Surface and Center Temperatures of a Sample with Time

Time (min)	T_s (°C)	T_c (°C)
0	25.0	25.0
2	30.6	26.6
4	35.2	28.2
6	39.7	30.7
8	45.3	36.1
10	49.8	40.7
12	54.8	45.9
14	60.0	51.2

P.3.8.1. A cylindrical container in which the sample was placed has a diameter of 3.6 cm. The center temperature (T_c) and surface temperature (T_s) were recorded at 2-min intervals until the surface temperature of the sample reached to 60°C. Calculate the thermal diffusivity of the sample.

REFERENCES

Abd Karim, A., Norziah, M.H., & Seow, C.C. (2000). Methods for the study of starch retrogradation. *Food Chemistry, 71*, 9–36.

Abramowitz, M., & Stegun, I.A. (1964). *Handbook of Mathematical Functions with Formulas, Graphs and Mathematical Tables.* NABS Applied Mathematics Series 55. Washington, DC: US Department of Commerce.

Bennett, A.H., Chace, W.G., & Cubbedge, R.H. (1962). Estimating thermal conductivity of fruit and vegetable components—the Fitch method. *ASHRAE Journal, 4,* 80–85.

Buhri, A.B., & Singh, R.P. (1993). Measurement of food thermal conductivity using differential scanning calorimetry. *Journal of Food Science, 58*, 1145–1147.

Carslaw, H.S., & Jaeger, J.C. (1959). *Conduction of Heat in Solids*, 2nd ed. Oxford, UK: Oxford University Press.

Carson, J.K., Lovatt, S.J., Tanner, D.J., & Cleland, A.C. (2004). Experimental measurements of the effective thermal conductivity of a pseudo-porous food analogue over a range of porosities and mean pore sizes. *Journal of Food Engineering, 63*, 87–95.

Carson, J.K., Lovatt, S.J., Tanner, D.J., & Cleland, A.C. (2005). Predicting the effective thermal conductivity of unfrozen, porous foods. *Journal of Food Engineering,* (http://dx.doi.org/10.1016/j.jfoodeng.2005.04.021).

Choi, Y., & Okos, M. R. (1986). Effects of temperature and composition on the thermal properties of foods. In Le M. Maguer, & P. Jelen (Eds.), *Food Engineering and Process Applications*, Vol. 1: *Transport Phenomena* (pp. 93–101). New York: Elsevier.

Dickerson, R.W. (1965). An apparatus for measurement of thermal diffusivity of foods. *Food Technology, 19*, 198–204.

Drouzas, A.E., & Saravacos, G.D. (1988). Effective thermal conductivity of granular starch materials. *Journal of Food Science, 53*, 1795–1798.

Eucken, A. (1940). Allgemeine Gesetzmassig keiten für das Wärmeleitvermogen veschiedener Stoffarten und Aggregatzustande. *Forschung Gebeite Ingenieur (Ausgabe A), 11*, 6.

Fitch, D.L., (1935). A new thermal conductivity apparatus. *American Physics Teacher, 3*, 135–136.

Gordon, C., & Thorne, S. (1990). Determination of the thermal diffusivity of foods from temperature measurements during cooling. *Journal of Food Engineering, 11*, 133–145.

Haas, E., & Felsenstein, G. (1978). *Methods used to the thermal properties of fruits and vegetables*, Special publication No. 103, Division of Scientific Publications, Bet Dagan, Israel: The Volcani Center.

Hamdami, N., Monteau, J., & Le Bail, A. (2003). Effective thermal conductivity of a high porosity model food at above and subfreezing temperatures. *International Journal of Refrigeration, 26*, 809–816.

Heissler, M.P. (1947). Temperature charts for induction and constant temperature heating. *A.S.M.E. Transactions, 69*, 227–236.

Heldman, D.R. (1975). *Food Process Engineering*. Westport, CT: AVI.

Hooper, F.C., & Lepper, F.R. (1950). Transient heat flow apparatus for the determination of thermal conductivities. *Transactions of the ASHRAE, 56*, 309–324.

Ingersoll, L.R. Zobel, O.J., & Ingersoll, A.C. (1954). *Heat Conduction with Engineering and Geological Applications*. New York: McGraw-Hill.

Kopelman, I.J. (1966). *Transient Heat Transfer and Thermal Properties in Food Systems*. PhD dissertation, Michigan State University, East Lansing, MI.

Krokida, M.K., Panagiotou, N.M., Maroulis, Z.B., & Saravacos, G.D. (2001). Thermal conductivity: Literature data compilation for foodstuffs. *International Journal of Food Properties, 4*, 111–137.

Loeb, L.B. (1965). *The Kinetic Theory of Gases*, 3rd ed. New York: Dover.

Luikov, A.V. (1964). Heat and mass transfer in capillary porous bodies. *Advanced Heat Transfer, 1*, 134.

Lund, D. (1984). Influence of time, temperature, moisture, ingredients and processing conditions of starch gelatinization. *CRC Critical Reviews in Food Science and Nutrition, 20*, 249–273.

Maroulis, Z.B., Saravacos, G.D., Krokida, M.K., & Panagiotou, N.M. (2002). Thermal conductivity prediction for foodstuffs: Effect of moisture content and temperature, *International Journal of Food Properties, 5*, 231–245.

Maxwell, J.C. (1904). *Treatise on Electricity and Magnetism*, 3rd ed. Oxford, UK: Clarenden Press.

McCurry, T.A. (1968). The development of a numerical technique for determining thermal diffusivity utilizing data obtained through the use of a refined line-source method. Advanced Project Report, Auburn: Auburn University.

Mohsenin, N.N. (1980). *Thermal Properties of Foods and Agricultural Materials*. New York: Gordon and Breach.

Moreira, R.G, Palau, J., Sweat, V.E., & Sun, X. (1995). Thermal and physical properties of tortilla chips as a function of frying time. *Journal of Food Processing and Preservation, 19*, 175–189.

Murakami, E.G., & Okos, M.R. (1989). Measurement and prediction of thermal properties of foods. In R.P. Singh, & A.G. Medina (Eds.), *Food Properties and Computer-Aided Engineering of Food Processing Systems* (pp. 3–48). Dordrecht: Kluwer Academic.

Murakami, E.G., Sweat, V.E., Sastry, S.K., & Kolbe, E. (1996a). Analysis of various design and operating parameters of thermal conductivity probe. *Journal of Food Engineering, 30*, 209–225.

Murakami, E.G., Sweat, V.E., Sastry, S.K., Kolbe, E., Hayakawa, K., & Datta, A. (1996b). Recommended design parameters for thermal conductivity probes for non-frozen food materials. *Journal of Food Engineering, 27*, 109–123.

Nix, G.H., Lowery, G.W., Vachon, R.I., & Tanger, G.E. (1967). Direct determination of thermal diffusivity and conductivity with a refined line-source technique. In G. Heller (Ed.), *Progress in Aeronautics and Astronautics: Thermophysics of Spacecraft and Planetary Bodies* (Vol. 20, pp. 865–878). New York: Academic Press.

Ohlsson, T. (1983). The measurement of thermal properties. In R. Jowitt (Ed.), *Physical Properties of Foods*. London: Applied Science.

Olson, F.C.W., & Schultz, O.T. (1942). Temperature in solids during heating and cooling, *Industrial & Engineering Chemistry, 34*, 874.

Rahman, M.S. (1992). Thermal conductivity of four food materials as a single function of porosity and water content. *Journal of Food Engineering, 15*, 261–268.

Rahman, M.S. (1993). Specific heat of selected fresh seafood. *Journal of Food Science, 58*, 522–524.

Rahman, M.S. (1995). *Food Properties Handbook.* New York: CRC Press.

Rahman, M.S., & Chen, X.D. (1995). A general form of thermal conductivity equation for an apple sample during drying. *Drying Technology, 13*, 2153–2165.

Rahman, M.S., Chen, X.D., & Perera, C.O. (1997). An improved thermal conductivity prediction model for fruits and vegetables as a function of temperature, water content and porosity. *Journal of Food Engineering, 31*, 163–170.

Riegel, C.A. (1992). Thermodynamics of moist air. In A.F.C. Bridger (Ed.), *Fundamentals of Atmospheric Dynamics and Thermodynamics.* Singapore: World Scientific.

Sahin, S., Sastry, S.K., & Bayındırlı, L. (1999). Effective thermal conductivity of potato during frying: Measurement and modelling. *International Journal of Food Properties, 2*, 151–161.

Saravacos, G.D., & Maroulis, Z.B. (2001). *Transport Properties of Foods.* New York: Marcel Dekker.

Sastry, S.K., & Cornelius, B.D. (2002). *Aseptic Processing of Foods Containing Solid Particulates.* New York: John Wiley & Sons.

Sharity-Nissar, M., Hozawa, M., & Tsukuda, T. (2000). Development of probe for thermal conductivity measurement of food materials under heated and pressurized conditions. *Journal of Food Engineering, 43*, 133–139.

Siebel, E. (1892). Specific heats of various products. *Ice and Refrigeration, 2*, 256–257.

Singh, R.P. (1995). Thermal properties of frozen foods. In M.A. Rao & S.S.H. Rizvi (Eds.), *Engineering Properties of Foods*, 2nd ed. (pp. 139–167). New York: Marcel Dekker.

Solomons, C., & Gummings, J.P. (1964). Dynamic adiabatic calorimeter: An improved calorimetric apparatus. *Review of Scientific Instruments, 35*, 307–310.

Sweat, V.E. (1995). Thermal properties of foods. In M. A. Rao & S. S. H. Rizvi (Eds.), *Engineering Properties of Foods*, 2nd ed. (pp. 99–138). New York: Marcel Dekker.

Sweat, V.E., & Haugh, C.G. (1974). A thermal conductivity probe for small food samples. *Transactions of the ASAE, 17*, 56–58.

Takegoshi, E., Imura, S., Hirasawa, Y., & Takenaka, T. (1982). A method of measuring the thermal conductivity of solid materials by transient hot wire method of comparison. *Bulletin of the JSME, 25*, 395–402.

Tong, C.H., & Sheen, S. (1992). Heat flux sensors to measure effective thermal conductivity of multilayered plastic containers. *Journal of Food Processing and Preservation, 16*, 233–237.

Van der Held, E.F.M., & van Drunen, F.G. (1949). A method of measuring the thermal conductivity of liquids. *Physica, 15*, 865–881.

Vos, B.H. (1955). Measurement of thermal conductivity by a non-steady-state method, *Applied Scientific Research, 5*, 425.

Voudouris, N., & Hayakawa, K. (1994). Simultaneous determination of thermal conductivity and diffusivity of foods using a point heat source probe: a theoretical analysis. *Lebensmittel Wissenschaft und Technologie, 27*, 522–532.

West, E.D., & Ginnings, D.C. (1958). An adiabatic calorimeter for the range 30°C to 500°C. *Journal of Research of the National Bureau of Standards, 60* (4), 309–316.

Zuritz, C.A., Sastry, S.K., Mccoy, S.C., Murakami, E.G., & Blaisdell, J.L. (1989). A modified Fitch device for measuring the thermal conductivity of small food particles. *Transactions of the ASAE, 32*, 711–717.

CHAPTER 4

Electromagnetic Properties

SUMMARY

This chapter gives information about color and dielectric properties of foods. First, color measurement methods and color order systems are discussed. Then, principles and measurement methods of dielectric properties are summarized. In addition, recent studies on dielectric properties are discussed.

Selective absorption of different amounts of wavelength within the visible region determines the color of a food material. Color is an important physical property which determines the acceptance of a food by consumer. A variety of systems have been developed to describe colors. Color order systems are three-dimensional arrangements of color according to appearance. There are five color order systems used for foods: Munsell, CIE, CIE $L^*a^*b^*$ (CIELAB), Hunter Lab, and Lovibond.

Dielectric properties of foods are important because they show microwave or radiofrequency heating ability of a product. They can also be used in assessment of food quality. Dielectric properties are the dielectric constant and dielectric loss factor. The dielectric constant is the ability of a material to store microwave energy and dielectric loss factor is the ability of a material to dissipate microwave energy into heat. Dielectric properties are functions of moisture content, temperature, and composition of the material. The increase in moisture content increases dipolar rotation so it increases both the dielectric constant and loss factor. The effect of bound water on dielectric properties becomes insignificant in the presence of free water. The temperature dependence of dielectric properties depends on the presence of bound and free moisture content and salt content of foods.

4.1 INTERACTION OF OBJECTS WITH LIGHT

When electromagnetic radiation strikes an object, the resulting interaction is affected by the properties of an object such as color, physical damage, and presence of foreign material on the surface. Different types of electromagnetic radiation can be used for quality control of foods. For example, near-infrared radiation can be used for measuring moisture content, and internal defects can be detected by X-rays.

Electromagnetic radiation is transmitted in the form of waves and it can be classified according to wavelength and frequency. The electromagnetic spectrum is shown in Fig. 4.1.

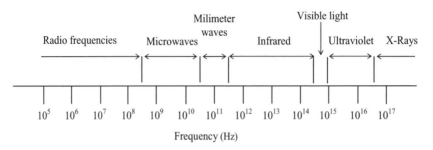

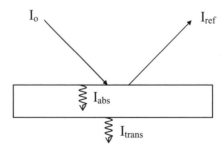

Figure 4.1 Electromagnetic spectrum.

Electromagnetic waves travel at the speed of light and are characterized by their frequency (f) and wavelength (λ). In a medium, these two properties are related by:

$$c = \lambda f \tag{4.1}$$

where c is the speed of light in vacuum (3.0×10^8 m/s).

Radiation can exhibit properties of both waves and particles. Visible light acts as if it is carried in discrete units called photons. Each photon has an energy, E, that can be calculated by:

$$E = hf \tag{4.2}$$

where h is Planck's constant (6.626×10^{-34} J·s).

When radiation of a specific wavelength strikes an object, it may be reflected, transmitted, or absorbed (Fig. 4.2). The relative proportion of these types of radiation determines the appearance of the object. A material is said to be transparent when the light impinging on it passes through with a minimum reflection and absorption. The opposite of transparency is opacity. That is, an object that does not allow any transmission of light but absorbs and/or reflects all the light striking is termed opaque. In opaque surfaces, certain specific wavelengths of light are absorbed and the others are reflected. As a result, color is formed. If all the visible light is absorbed, the object appears black. If both reflection and transmission occur, the material is said to be translucent.

If I_0 is the radiant energy striking the object and I_{ref} is the amount of energy reflected from the object (Fig. 4.3), the total reflectance R is defined as:

$$R = I_{ref}/I_0 \tag{4.3}$$

Figure 4.2 Effects of incident radiation.

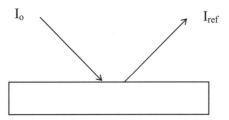

Figure 4.3   Reflection.

If I_0 is the incident energy, I_1 is the energy entering the object, I_2 is the energy striking the opposite face after transmission, and I_{out} is the energy leaving the opposite phase (Fig. 4.4), transmittance (T) and absorbance (A) are defined as:

$$T = I_{out}/I_0 \tag{4.4}$$

$$A = \log(I_1/I_2) \tag{4.5}$$

Optical density (OD) is a measure of the relative amount of incident energy transmitted through the object.

$$OD = \log(I_0/I_{out}) = \log(1/T) \tag{4.6}$$

There are two types of reflection: diffuse reflection and specular reflection. In diffuse reflection, radiation is reflected equally in all directions (Fig. 4.5). When a surface is rough, the incident light will bounce around and will rise to a greater amount of diffuse light. Opaque surfaces reflect light diffusely.

Specular reflection is highly directional instead of diffuse. The angle of reflection is equal to the angle of incidence of the radiation beam (Fig. 4.6). It is mainly responsible for the glossy or shiny

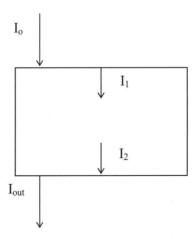

Figure 4.4   Transmission.

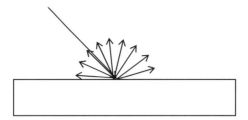

Figure 4.5   Diffuse reflection.

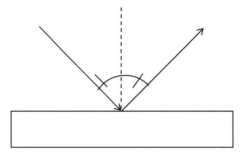

Figure 4.6   Specular reflection.

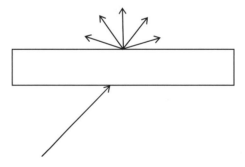

Figure 4.7   Diffuse transmission.

(mirror-like) appearance of the material. If the object is smooth, such as a mirror, most of the light will be reflected in this way.

There are two different types of transmission: diffuse transmission and rectilinear transmission. Diffuse transmission occurs when light penetrates an object, scatters, and emerges diffusely on the other side. Transmitted light leaves the object surface in all directions. It is seen visually as cloudiness, haze, or translucency (Fig. 4.7).

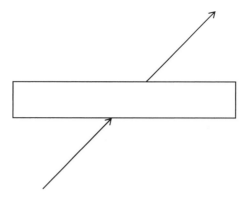

Figure 4.8 Rectilinear transmission.

Rectilinear transmission refers to the light passing through an object without diffusion. Rectilinear transmission measurements are widely used in the chemical analysis and color measurements of liquids (Fig. 4.8).

When light passes through one medium to another, the speed of light changes. This causes light to bend when it enters a different medium unless it hits directly perpendicular to the boundary (Fig. 4.9). This phenomenon is known as refraction. The ratio of speed of light in one medium (e.g., air) to the speed of light in another is called the refractive index of the material. This property is used in food analysis applications such as determining the alcohol content or sucrose concentration in percentage or in Brix.

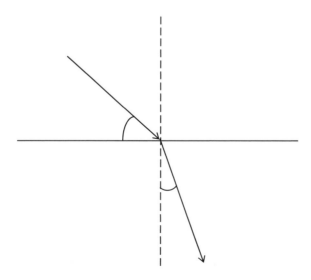

Figure 4.9 Refraction of light.

4.2   COLOR

Color is one of the important quality attributes in foods. Although it does not necessarily reflect nutritional, flavor, or functional values, it determines the acceptability of a product by consumers.

Sometimes, instead of chemical analysis, color measurement may be used if a correlation is present between the presence of the colored component and the chemical in the food since color measurement is simpler and quicker than chemical analysis. For example, total carotenoid content of squash can be determined from color measurements without performing a chemical analysis because there is a correlation between total carotenoid content and color for squash (Francis, 1962).

It may be desirable to follow the changes in color of a product during storage, maturation, processing, and so forth. Color is often used to determine the ripeness of fruits. Color of potato chips is largely controlled by the reducing sugar content, storage conditions of the potatoes, and subsequent processing. Color of flour reflects the amount of bran. In addition, freshly milled flour is yellow because of the presence of xanthophylls.

Color is a perceptual phenomenon that depends on the observer and the conditions in which the color is observed. It is a characteristic of light, which is measurable in terms of intensity and wavelength. Color of a material becomes visible only when light from a luminous object or source illuminates or strikes the surface.

Light is defined as visually evaluated radiant energy having a frequency from about 3.9×10^{14} Hz to 7.9×10^{14} Hz in the electromagnetic spectrum (Fig. 4.1). Light of different wavelengths is perceived as having different colors. Many light sources emit electromagnetic radiation that is relatively balanced in all of the wavelengths contained in the visible region. Therefore, light appears white to the human eye. However, when light interacts with matter, only certain wavelengths within the visible region may be transmitted or reflected. The resulting radiations at different wavelengths are perceived by the human eye as different colors, and some wavelengths are visibly more intense than others. That is, the color arises from the presence of light in greater intensities at some wavelengths than at the others.

The selective absorption of different amounts of the wavelengths within the visible region determines the color of the object. Wavelengths not absorbed but reflected by or transmitted through an object are visible to observers. For example, a blue object reflects the blue light spectrum but absorbs red, orange, yellow, green, and violet. If all radiant energy in the visible region is reflected from an opaque surface, the object appears white. If it is almost completely absorbed, the object is black. When absorption is the dominant process, the resulting colors are not intense.

Physically, the color of an object is measured and represented by spectrophotometric curves, which are plots of fractions of incident light (reflected or transmitted) as a function of wavelength throughout the visible spectrum (Fig. 4.10).

An object, a light source or an illuminant, and an observer are required for the presence of color. Light sources and illuminants are often confused. A light source can be turned on and off and can be used to view an object. An illuminant, on the other hand, is a mathematical description of a light source.

For detection of differences in color under diffuse illumination, both natural daylight and artificial simulated daylight are commonly used. A window facing north that is free of direct sunshine is the natural illuminant normally employed for visual color examination. However, natural daylight varies greatly in spectral quality with direction of view, time of day and year, weather, and geographical location. Therefore, simulated daylight is commonly used in industrial testing. Artificial light sources can be standardized and remain stable in quality. The Commission Internationale de l'Eclairage (CIE)

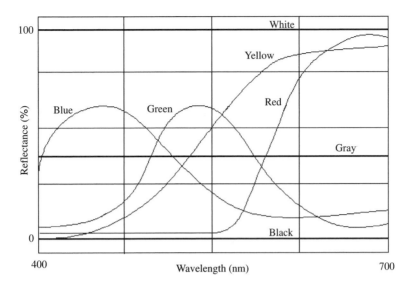

Figure 4.10   Spectrophotometric curves.

(The International Commission on Illumination) recommended three light sources reproducible in the laboratory in 1931. Illuminant A defines light typical of that from an incandescent lamp, illuminant B represents direct sunlight, and illuminant C represents average daylight from the total sky. Based on measurements of daylight, CIE recommended a series of illuminants D in 1966 to represent daylight. These illuminants represent daylight more completely and accurately than illuminants B and C do. In addition, they are defined for complete series of yellow to blue color temperatures. The D illuminants are usually identified by the first two digits of their color temperature.

Incandescence and luminescence are two main ways of creating light. Incandescence is the light from heat energy. Heating the source of a light bulb to a sufficiently high temperature will cause it to glow. The glow of stars and sun is via incandescence. Luminescence is the light from other sources of energy independent of heating. Therefore, it is also known as cold light. It can be generated at room or even lower temperatures. Quantum physics explains luminescence as movements of electrons from their ground state (lowest energy level) to a state of high energy. The electron gives back the energy in the form of a photon of light when it returns to its ground state. If the time interval between the two steps is short (a few microseconds), the process is called fluorescence. If the interval is long (hours), the process is called phosphorescence. Fluorescence is used to detect the presence of *Aspergillus*, which produces aflatoxin.

The human eye is the defining observer of color. The eye can determine approximately 10 million colors, which is not possible for the instruments. People who believe that the eye is the most important observer of color argue that judgments of color can be made purely by visual matching with color cards. However, everyone's perception of color differs. Color match can become highly subjective as color vision varies widely from person to person depending on age, gender, inherited traits, and mood. Therefore, instrumental methods for color measurement have been developed.

4.2.1   Color Measuring Equipments

Color measuring instruments are categorized into two types: spectrophotometers and colorimeters.

4.2.1.1   *Spectrophotometers*

Early instrumental methods for color measurement were based on transmission or reflection spectrophotometry (Billmeyer & Saltzman, 1981). In spectrophotometers, three projectors each with red, green, or blue filters in front of the lens are required. Red, green, or blue light beams are focused on a screen such that they overlap over half a circle. The other half is illuminated by another projector or by spectrally pure light from a prism or grating. The observer can see both halves of the circle on the screen simultaneously. Each projector is equipped with a rheostat to vary the amount of light from each of the red, green, and blue sources. The observer can determine the amounts of red, green, and blue required to match almost any spectral color by varying the amount of light. Spectral color can be defined in terms of the amounts of red, green, and blue (Francis, 1983).

Vision scientists have created a special set of mathematical functions X, Y, and Z to replace red, green, and blue lights, respectively. The color matching functions for X, Y, and Z lights are all positive numbers and are labeled as $\bar{x}$, $\bar{y}$, $\bar{z}$. The color can be matched using the appropriate amounts of X, Y, and Z light. The amount of X, Y, and Z light required to match a color are called the color's tristimulus.

The red, green, and blue data for spectral colors are taken and transformed into X, Y, and Z coordinates. Then, the response of the human eye against wavelength was plotted (Fig. 4.11). These standardized curves are called the CIE $\bar{x}$, $\bar{y}$, $\bar{z}$ standard observer curves.

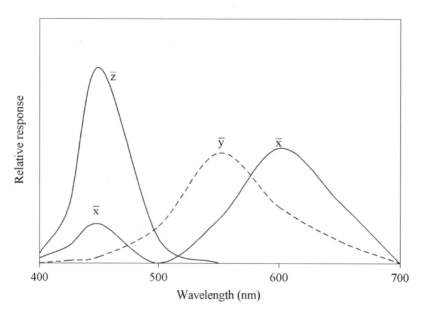

Figure 4.11   Standard observer curves.

Using the data in Fig. 4.11, color can be calculated from a reflection or a transmission spectrum. The sample spectrum (R) is multiplied by the spectrum of the light source (E) and the area under the curve is integrated in terms of the $\bar{x}$, $\bar{y}$, $\bar{z}$ standard observer curves (Francis, 2005). This can be described mathematically by the integral equations:

$$X = \int_{380}^{750} RE\bar{x}dx \tag{4.7}$$

$$Y = \int_{380}^{750} RE\bar{y}dy \tag{4.8}$$

$$Z = \int_{380}^{750} RE\bar{z}dz \tag{4.9}$$

In a spectrophotometer, light is usually split into a spectrum by a prism or a diffraction grating before each wavelength band is selected for measurement. Instruments have also been developed in which narrow bands are selected by interference filters. The spectral resolution of the instrument depends on the narrowness of the bands utilized for each successive measurement. Spectrophotometers contain monochromators and photoiodes that measure the reflectance curve of a product's color every 10 nm or less. The analysis generates typically 30 or more data points with which a precise color composition can be calculated.

Spectrophotometers measure the reflectance for each wavelength and allow calculation of tristimulus values. The advantage of spectrophotometers over tristimulus colorimetry is that adequate information is obtained to calculate color values for any illuminant and metamerism which is the difference in color in different lighting and at different angles is automatically detected. However, high-quality spectrophotometers are very expensive and measurements take longer.

4.2.1.2 Colorimeters

Before the 1950s, calculation of X, Y, Z data from spectra was very common but it required computational work which led to the development of electronic integration. Tristimulus colorimeters were developed since spectrophotometer integration was expensive. A tristimulus colorimeter has three main components:

1. Source of illumination
2. Combination of filters used to modify the energy distribution of the incident/reflected light
3. Photoelectric detector that converts the reflected light into an electrical output

Each color has a fingerprint reflectance pattern in spectrum. The colorimeter measures color through three wide-band filters corresponding to the spectral sensitivity curves. Measurements made on a tristimulus colorimeter are normally comparative. It is necessary to use calibrated standards of similar colors to the materials to be measured to achieve the most accurate measurements.

Each color has its own tristimulus values that distinguish it from any other color. These values can be measured to determine if a color match is accurate. They can also be used to determine the direction and amount of any color difference.

4.2.2 Color Order Systems

A variety of systems have been developed to describe colors. Color order systems are three-dimensional arrangements of color according to appearance. Each color has a notation relating to its position in the arrangement.

4.2.2.1 Munsell Color System

The Munsell Color System was developed in 1898 by an American artist and teacher, Albert Munsell (Marcus, 1998). Thousands of colors could be described by using hue, value (lightness), and chroma (saturation) in this color system (Fig. 4.12).

Hue is a quality by which one color is distinguished from another. It is the attribute corresponding to whether the object is red, orange, yellow, blue, or violet. This perception of color results from differences in absorption of radiant energy at various wavelengths. If the shorter wavelengths of 425 to 490 nm are reflected to a greater extent than the other wavelengths, the color is described as blue. Maximum reflection in the medium wavelength range results in green or yellow color. Maximum reflectance at the longer wavelengths of 640 to 740 nm indicates red objects. Hue is defined on the Munsell color tree in the circumferential direction by five principal and five intermediate hues. The letters used to designate the main hues are R (red), Y (yellow), G (green), B (blue), P (purple) and their intermediates YR (yellow-red), GY (green-yellow), BG (blue-green), PB (purple-blue), and RP (red-purple). Each is at the midpoint of a scale from 1 to 10. One may start at 1R: 1R, 2R, . . . , 5R (or R), . . . , 10R, 1YR, . . . , 5YR (YR), . . . , 10 YR, etc. and so on around the circle to 10RP and 1R again. A hue scale of 1 to 100 is also encountered.

Munsell defined value as the quality by which light colors can be differentiated from the dark ones. Value is a neutral axis that refers to the gray level of the color ranging from white to black (Fig. 4.12). It is also called lightness, luminous intensity, and sometimes brightness. It describes the relationship between reflected and absorbed light without regard to specific wavelength. The value scale contains 10 steps in the vertical direction, from black (0) to white (10) in the Munsell system.

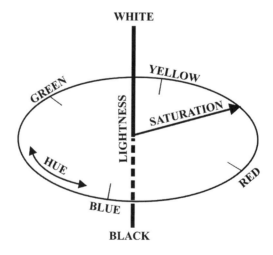

Figure 4.12   Three-dimensional Munsell color system.

"Chroma" is the quality that distinguishes a pure hue from a gray shade. The chroma (saturation or purity) axis extends from the value (lightness) axis toward the pure hue. A pastel tint has low saturation while pure color is said to have high saturation. Two objects having the same lightness and hue can be differentiated from their saturation. In the Munsell system, chroma is defined by 10 or more steps (up to 16 steps) in the radial direction. It is a measure of the difference from a gray of the same lightness.

Munsell color notations are always written in the same sequence as hue value/chroma. Thus, a color represented by 6.5R 2.6/2.2 indicates a product with red hue of 6.5, a value of 2.6, and a chroma of 2.2.

In the Munsell system, three or four overlapping discs are used, one each for hue and chroma and one or two to provide adjustment for value. The discs are adjusted until the color obtaining by rapid spinning of the discs matches the color of the tested object.

Computer programs are available that allow the calculation of Munsell notations from a sample's X, Y, and Z tristimulus values.

4.2.2.2   CIE Color System

One of the well known systems to describe colors is the CIE color system developed by the International Commission on Illumination in 1931. This is a trichromatic system, that is, any color can be matched by a suitable mix of the three primary colors—red, green, and blue which are represented by X, Y, and Z, respectively.

This system uses the chromaticity diagram to designate various colors. When determining the specification for a color, the reflectance and transmittance at each wavelength are measured. These values are weighted by functions that represent the relative intensities of reflectance at various wavelengths which would be perceived as blue, green, and red by a standard observer. The application of the weighting to a reflectance curve gives the tristimulus values, which are denoted by the capital letters X, Y, and Z. These values are then used to calculate the chromaticity coordinates, designated by lowercase letters x (red), y (green), and z (blue). The value for x can be calculated by:

$$x = \frac{X}{X + Y + Z} \tag{4.10}$$

The values for y and z can be calculated by replacing X with Y and Z, respectively, in the numerator. Since the sum of the chromaticity coordinates (x, y, and z) is always unity, the values x and y alone can be used to describe a color. When x and y are plotted, a chromaticity diagram is obtained (Fig. 4.13). The third dimension of lightness is defined by the Y tristimulus value. Color is defined by its chromaticity coordinates x and y and its lightness, the tristimulus Y value. All real colors lie within the horseshoe-shaped locus marked with the wavelengths of the spectrum colors. The color is read from the diagram. Once the point has been located on the diagram, the hue (dominant wavelength) and the purity (percent saturation) of the color are determined.

4.2.2.3   CIE $L^*a^*b^*$ (CIELAB) Color Spaces

The CIELAB color measurement method was developed in 1976 and offers more advantages over the system developed in 1931. It is more uniform and based on more useful and accepted colors describing a theory of opposing colors.

The location of any color in the CIELAB color space is determined by its color coordinates: L^*, a^*, and b^*. L^* represents the difference between light ($L^* = 100$) and dark ($L^* = 0$). The component a^* represents the difference between green ($-a^*$) and red ($+a^*$) and component b^* represents the

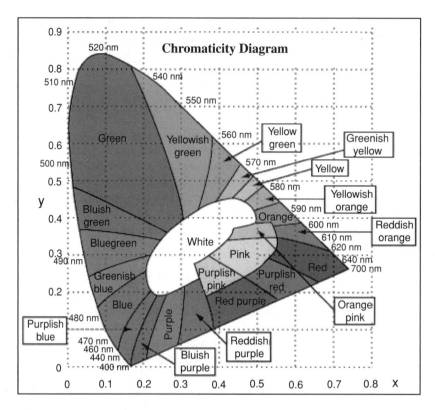

Figure 4.13   Chromaticity diagram of the CIE system.

difference between blue $(-b^*)$ and yellow $(+b^*)$. If the L^*, a^*, and b^* coordinates are known, then the color is not only described, but also located in the space.

CIELAB L^*, a^*, and b^* coordinates are calculated from the tristimulus values according to the following equations (Marcus, 1998):

$$L^* = 116 f\left(Y/Y_n\right) - 16 \tag{4.11}$$

$$a^* = 500\left[f\left(X/X_n\right) - f\left(Y/Y_n\right)\right] \tag{4.12}$$

$$b^* = 200\left[f\left(Y/Y_n\right) - f\left(Z/Z_n\right)\right] \tag{4.13}$$

in which the subscript n refers to the tristimulus values of the perfect diffuser for the given illuminant and standard observer.

$$f\left(\frac{X}{X_n}\right) = \left(\frac{X}{X_n}\right)^{1/3} \text{ for } \left(\frac{X}{X_n}\right) > 0.008856 \text{ and}$$

$$f\left(\frac{X}{X_n}\right) = 7.787\left(\frac{X}{X_n}\right) + \frac{16}{116} \text{ for } \left(\frac{X}{X_n}\right) \leq 0.008856:$$

The same color equations can be used for Y and Z by replacing X.

Colors can also be located using cylindrical coordinates in CIELAB color space. The cylindrical version of L^*, a^*, b^* system is termed as the CIE $L^*C^*h^*$ color space and it resembles the Munsell color order system, corresponding to the perceptual attributes of lightness (L^*), chroma (C^*), and hue (h^*) (McLaren, 1976). L^* is the lightness coordinate as in $L^*a^*b^*$; C^* is the chroma coordinate, which is perpendicular distance from the lightness; and h^* is the hue angle expressed in degrees with $0°$ being a location on the $+a^*$ axis, continuing to $90°$ for the $+b^*$ axis, $180°$ for $-a^*$, $270°$ for $-b^*$, and back to $360° = 0°$. Mathematically, chroma (C^*) and hue angle (h^*) are defined as:

$$C^* = \left[a^{*2} + b^{*2} \right]^{1/2} \tag{4.14}$$

$$h^* = \arctan \left[\frac{b^*}{a^*} \right] \tag{4.15}$$

Many CIE system users prefer the $L^*C^*h^*$ space for specifying a color, since the concept of hue and chroma agrees well with visual experience.

4.2.2.4   Hunter Lab Color Space

This system is based on L, a, and b measurements. The L value represents lightness and changes from 0 (black) to 100 (white). The a value changes from $-a$ (greenness) to $+a$ (redness) while the b value is from $-b$ (blueness) to $+b$ (yellowness). Like the CIE system, the Hunter scale is also derived from X, Y, Z values.

In the human eye, there is an intermediate signal switching stage between the light receptors in the retina and the optic nerve taking color signals to the brain. In this switching stage, red responses are compared with green to generate a red-to-green color dimension. The green (or red and green together) response is compared in a similar manner with the blue to generate a yellow to blue color dimension. These two dimensions are associated with the symbols a and b, respectively. The a value is a function of X and Y, the b value of Z and Y. The necessary third dimension, L for lightness, is a nonlinear function such as the square or cube root of Y which is percent reflectance (or transmittance). The variables L, a, and b for the Hunter Lab system are defined in terms of tristimulus values, X, Y, and Z, as in Eqs. (4.16) to (4.18) for standard daylight.

$$L = 10.0\sqrt{Y} \tag{4.16}$$

$$a = \frac{17.5(1.02X - Y)}{\sqrt{Y}} \tag{4.17}$$

$$b = \frac{7.0(Y - 0.847Z)}{\sqrt{Y}} \tag{4.18}$$

The dimensions of the Hunter L, a, b system are shown in Fig. 4.14. The Hunter L (lightness) values are directly comparable to the Y of the CIE system or value of the Munsell system. Determining a and b values provides information equivalent to that of determining the hue and chroma dimensions of the Munsell system.

4.2.2.5   Lovibond System

In the CIE system, X, Y, and Z are added in certain proportions to match a given color. The Munsell space is developed by an entirely different principle, namely, a systematic sampling of the color space, but when the spinning disc technique is used, it creates in principle an additive colorimeter. The Lovibond scale is based on the opposite principle. It starts with white and then by the use of red,

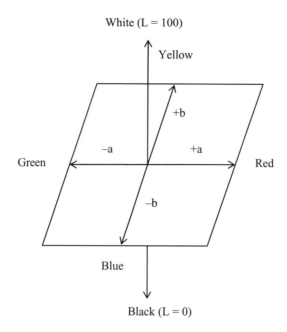

Figure 4.14   Hunter L, a, b color system.

---

yellow, and blue filters, colors are subtracted from the original white to achieve the desired match with the sample (MacKinney & Little, 1962).

Lovibond is a standard scale for the measurement of beer color. The Lovibond tintometer is a visual colorimeter used widely in the oil industry. Lovibond measurement for color is also used in the honey industry. The instrument has a set of permanent glass color filters in the three primary colors: red, yellow, and blue. The sample is placed in a glass cell and the filters are introduced into the optical system until a color match is obtained under specified conditions of illumination and viewing. The color of the sample is measured with transmitted light.

The Lovibond measurements of different colors are given in Table 4.1.

4.2.3   Color Differences

CIE $L^*a^*b^*$ (CIELAB) total color difference (ΔE^*) is the distance between the color locations in CIE space. This distance can be expressed as:

$$\Delta E^* = \left[\left(\Delta L^* \right)^2 + \left(\Delta a^* \right)^2 + \left(\Delta b^* \right)^2 \right]^{1/2} \tag{4.19}$$

where ΔL^* is the lightness difference:

$$L^*_{\text{sample}} - L^*_{\text{standard}}$$

Δa^* is the red/green difference:

$$a^*_{\text{sample}} - a^*_{\text{standard}},$$

and Δb^* is the yellow/blue difference: $b^*_{\text{sample}} - b^*_{\text{standard}}.$

Table 4.1 The Lovibond Measurements
of Different Colors

Color	Lovibond Measurement
Light yellow	2.0–3.0
Medium yellow	3.0–4.5
Deep straw/gold	4.5–6.0
Deep gold	6.0–7.5
Light amber	7.5–9.0
Copper	9.0–11.0
Red/brown	11–14
Light brown	14–17
Medium brown	17–20
Light black	20–25
Black	>25

ΔE^* is an equally weighted combination of the coordinate (L^*, a^*, b^*) differences. It represents the magnitude of the difference in color but does not indicate the direction of the color difference.

Color difference can also be described specifying L^*, C^*, and h^* coordinates as:

$$\Delta E^* = \left[\left(\Delta L^*\right)^2 + \left(\Delta C^*\right)^2 + \left(\Delta H^*\right)^2\right]^{1/2} \tag{4.20}$$

The chroma difference between sample and standard is given as:

$$\Delta C^* = C^*_{\text{sample}} - C^*_{\text{standard}} \tag{4.21}$$

The metric hue difference, ΔH^*, is not calculated by subtracting hue angles. It is expressed as:

$$\Delta H^* = \left[\left(\Delta a^*\right)^2 + \left(\Delta b^*\right)^2 - \left(\Delta C^*\right)^2\right]^{1/2} \tag{4.22}$$

There are various recent papers in which total color difference (ΔE^*) has been used in determining color change in different foods during processing and storage. The following are some examples of these papers, which are concerned with color change (ΔE value) during frying of chicken nuggets (Altunakar, Sahin, & Sumnu, 2004), frying of tofu (Baik & Mittal, 2003), deep-fat frying of french fries (Moyono, Rioseco, & Gonzalez, 2002), frying of doughnuts (Vélez-Ruiz & Sosa-Morales, 2003), halogen lamp-microwave combination baking of cakes (Sevimli, Sumnu, & Sahin, 2005), conventional, microwave and halogen lamp baking of bread (Keskin, Sumnu, & Sahin, 2004), thermal and nonthermal processing of apple cider (Choi & Nielsen, 2005), thermal processing of pineapple juice (Rattanathanalerk, Chiewchan, & Srichumpoung, 2005), drying of onions (Kumar, Hebbar, Sukumar, & Ramesh, 2005), drying of green table olives (Ongen, Sargin, Tetik, & Kose, 2005), microwave and infrared drying of carrot and garlic (Baysal, Icier, Ersus, & Yildiz, 2003), hot air and microwave drying of kiwi fruit (Maskan, 2001), and toasting of bread (Ramirez-Jimenez, Garcia-Villanova, & Guerra-Hernandez, 2001).

Color changes in foods are almost always three-dimensional but not all three dimensions may be of practical importance. The number of color parameters can be reduced by making a correlation between

Table E.4.1.1   Color Values of Potato Slices During Frying

Frying Time (min)	L^*	a^*	b^*
2.0	69.63	0.567	39.20
2.5	67.47	2.467	45.10
3.0	63.67	3.033	46.00

visual scores and color readings. The color parameter(s) that have a high correlation coefficient with the visual scores are selected for discussing the results.

Example 4.1. Colorimetric properties of potato slices during microwave frying in sunflower oil are studied in terms of a CIE scale. As a standard, a $BaSO_4$ plate with L^*, a^*, and b^* values of 96.9, 0.0, and 7.2, respectively was used. L^*, a^*, and b^* values of potato slices are given in Table E.4.1.1. Determine ΔE^* values of the potato slices during frying and discuss the results.

Solution:

Subtracting the standard color values from those of french fries, ΔL^*, Δa^*, and Δb^* values were determined and given in Table E.4.1.2:

$$\Delta L^* = L^*_{\text{sample}} - L^*_{\text{standard}}$$
$$\Delta a^* = a^*_{\text{sample}} - a^*_{\text{standard}}$$
$$\Delta b^* = b^*_{\text{sample}} - b^*_{\text{standard}}$$

Then, Eq. (4.19) is used to determine the ΔE^* value:

$$\Delta E^* = \left[\left(\Delta L^*\right)^2 + \left(\Delta a^*\right)^2 + \left(\Delta b^*\right)^2\right]^{1/2} \tag{4.19}$$

For frying time of 2 min:

$$\Delta E^* = \left[(-27.27)^2 + (0.567)^2 + (32)^2\right]^{1/2} = 42.05$$

Table E.4.1.2 ΔL^*, Δa^*, Δb^*, and ΔE^* Values for Potato Slices

Frying Time (min)	ΔL^*	Δa^*	Δb^*	ΔE^*
2.0	−27.27	0.567	32.00	42.05
2.5	−29.43	2.467	37.90	48.05
3.0	−33.23	3.033	38.80	51.17

For the other frying times, calculated ΔE^* values are given in Table E.4.1.2. As can be seen in this table as frying time increases, ΔE^* value increases showing that colors of potatoes become darker.

4.3 DIELECTRIC PROPERTIES OF FOODS

Electromagnetic heating such as microwave and radiofrequency (RF) heating are used in many processes such as reheating, precooking, tempering, baking, drying, pasteurization, and sterilization in industry and at home, including electromagnetic heating processes related to dielectric properties of the material. Since microwave heating is common in many food processes, determination of dielectric properties becomes significant to understand the heating profiles of foods in a microwave oven, and to develop equipment and microwaveable foods.

4.3.1 Basic Principles of Microwave Heating

Microwaves, which are electromagnetic waves, cover a spectrum of frequencies ranging from 300 MHz to 30 GHz. Microwaves, like light waves, are reflected by metallic objects, absorbed by dielectric materials, or transmitted from glass.

Although microwaves cover a wide range of frequencies, their use is restricted to some frequencies owing to the possibility of interference of microwaves with radar or other communication devices. The typical frequencies used in microwaves are 2450 MHz for home type ovens and 915 MHz for industrial use.

Absorption of microwave energy in the food involves primarily two mechanisms: ionic interaction and dipolar rotation.

4.3.1.1 Ionic Interaction (Ionic Conduction)

Figure 4.15a illustrates the mechanism of ionic conduction. Salt, a common molecule in foods, is composed of positive sodium and negative chloride ions in dissociated form. The net electric field in the oven will accelerate the particle in one direction and the oppositely charged particle in the other direction. If the accelerating particle collides with an adjacent particle, it will impart kinetic energy to it and set it into more agitated motion. As a result of agitation, the temperature of the particle increases. More agitated particles interact with their neighbors and transfer agitation or heat to them. This heat is then transferred to the other parts of the material.

4.3.1.2 Dipolar Rotation

The physically separated charges are called dipoles. Molecules with such separated charges are known as polar molecules. Molecules having a center of symmetry such as methane (CH_4) are nonpolar and exhibit zero dipole moment. Molecules such as water or gelatin are polar because they have no charge symmetry and exhibit strong dipole moments. Water in food is the primary component responsible for dipolar rotation.

If the water molecules are placed in an alternating electric field, they will experience a rotational force attempting to orient them in the direction of the field (Fig. 4.15b). As molecules attempt to orient themselves in the field direction, they collide randomly with their neighbors. When the field reverses

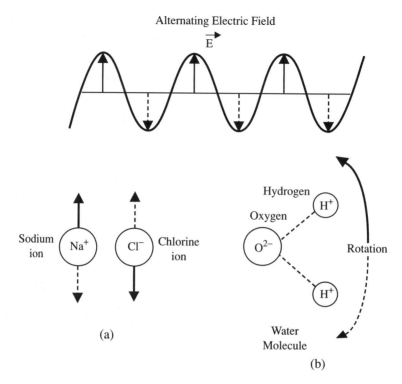

Figure 4.15   Mechanisms of microwave interaction with food. (a) Ionic; (b) dipolar.

its direction, they try to line up with the reversed direction and further collisions occur. This causes thermal agitation and heating takes place.

4.3.2   Definition of Dielectric Properties

Dielectric properties can be categorized into two: dielectric constant and dielectric loss factor. Dielectric constant (ε') is the ability of a material to store microwave energy and dielectric loss factor (ε'') is the ability of a material to dissipate microwave energy into heat. The parameter that measures microwave absorptivity is the loss factor. The values of dielectric constant and loss factor will play important roles in determining the interaction of microwaves with food.

The rate of heat generation per unit volume (Q) at a location inside the food during microwave heating can be characterized by Eq. (4.23):

$$Q = 2\pi f \varepsilon_0 \varepsilon'' E^2 \tag{4.23}$$

where f is the frequency, ε_0 is the dielectric constant of free space (8.854×10^{-12} F/m), ε'' is the dielectric loss factor, and E is the electric field.

As microwaves move through the slab at any point, the rate of heat generated per unit volume decreases. For materials having a high loss factor, the rate of heat generated decreases rapidly and microwave energy does not penetrate deeply. A parameter is necessary to indicate the distance that microwaves will penetrate into the material before it is reduced to a certain fraction of its initial value.

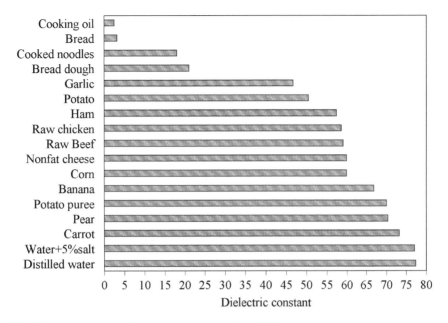

Figure 4.16   Dielectric constant of various food materials at 25°C.

This parameter is called power penetration depth (δ_p), which is defined as the depth at which power decreases to $1/e$ or (36.8%) of its original value. It depends on both dielectric constant and loss factor of food.

$$\delta_p = \frac{\lambda_0}{2\pi \sqrt{2\varepsilon'}} \left(\sqrt{1 + (\varepsilon''/\varepsilon')^2} - 1 \right)^{-\frac{1}{2}} \tag{4.24}$$

where λ_0 is wavelength of the microwave in free space.

Dielectric constant and loss factor of various food materials can be seen in Figs. 4.16 and 4.17, respectively. As can be seen in the figures, dielectric properties of cooking oil are very low because of its nonpolar characteristic. Dielectric properties of water and high-moisture-containing foods such as fruits, vegetables, and meat are high because of dipolar rotation. The highest loss factor is observed in the case of salt-containing foods such as ham.

Comprehensive reviews on dielectric properties provide good sources of experimental data for many foods (Datta, Sumnu, & Raghavan, 2005; Nelson & Datta, 2001; Venkatesh & Raghavan, 2004). Studies in recent years include determination of dielectric properties of ham (Sipahioglu, Barringer, & Taub, 2003a), turkey meat (Sipahioglu, Barringer, Taub, & Yang, 2003b), fruits and vegetables (Sipahioglu & Barringer, 2003), starch solutions (Piyasena, Ramaswamy, Awuah, & Defelice, 2003), glucose solutions (Liao, Raghavan, Meda, & Yaylayan, 2001: Liao, Raghavan, Dai, & Yaylayan, 2003), mashed potatoes (Regier, Housova, & Hoke, 2001; Guan, Cheng, Wang, & Tang, 2004), macaroni (Wang, Wig, Tang, & Hallberg, 2003), various food proteins (Bircan & Barringer, 2002a,b; Bircan, Barringer, & Mangino, 2001; Wang et al., 2003), oysters (Hu & Mallikarjunan, 2005), meat and meat ingredients (Lyng, Zhang, & Brunton, 2005), and meat batters (Zhang, Lyng, Brunton, Morgan, & McKenna, 2004).

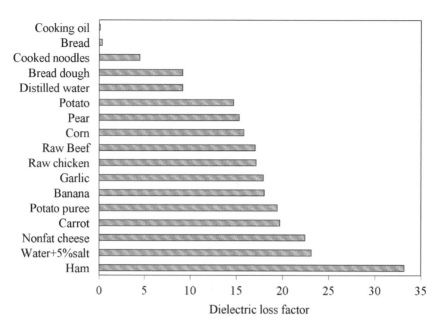

Figure 4.17   Dielectric loss factor of various food materials at 25°C.

Dielectric properties of foods depend on moisture content, temperature, and compositional properties of foods. They are also a function of the frequency of the oven. Information about the effects of frequency on dielectric properties can be found in the review of Datta, Sumnu, and Raghavan (2005) and Nelson and Datta (2001).

Example 4.2. Estimate the penetration depth of a chicken meat during processing in home type microwave oven. Chicken meat has a dielectric constant of 53.2 and dielectric loss factor of 18.1. Assume that dielectric properties are constant during heating.

Solution:

The frequency of a home type microwave oven is 2450 MHz.
Wavelength in free space is calculated as:

$$\lambda_0 = \frac{c}{f} = \frac{3 \times 10^8}{2450 \times 10^6} = 0.122 \text{ m}$$

Using Eq. (4.24):

$$\delta_p = \frac{\lambda_0}{2\pi \sqrt{2\varepsilon'}} \left(\sqrt{1 + (\varepsilon''/\varepsilon')^2} - 1 \right)^{-\frac{1}{2}} \tag{4.24}$$

$$\delta_p = \frac{0.122}{2\pi \sqrt{(2)(53.2)}} \left(\sqrt{1 + \left(\frac{18.1}{53.2}\right)^2} - 1 \right)^{-1/2} = 0.00794 \text{ m}$$

4.3.3   Effects of Moisture Content on Dielectric Properties

Liquid water is very polar and can easily absorb microwave energy based on the mechanism of dipolar rotation. The dielectric constant and loss factor of free water are predicted by Debye models and shown in Eqs. (4.25) and (4.26), respectively (Mudgett, 1995). Debye models are expressed in terms of wavelength and temperature-dependent parameters.

$$\varepsilon' = \frac{\varepsilon_s - \varepsilon_0}{1 + \left(\lambda_s / \lambda\right)^2} + \varepsilon_0 \tag{4.25}$$

$$\varepsilon'' = \frac{(\varepsilon_s - \varepsilon_0)\left(\lambda_s / \lambda\right)}{1 + \left(\lambda_s / \lambda\right)^2} \tag{4.26}$$

where ε_s is static dielectric constant, ε_0 is optical dielectric constant, λ is wavelength of water, and λ_s is critical wavelength of polar solvent.

Water can exist in either the free or bound state in food systems. Free water is found in capillaries but bound water is physically adsorbed to the surface of dry material. The dielectric loss factor is affected by the losses in free and bound water but since relaxation of bound water takes place below microwave frequencies, its effects are small in microwave processing (Calay, Newborough, Probert, & Calay, 1995). Figure 4.18 shows the variation of dielectric loss factor with moisture content. As can be seen in the figure, loss factor is constant in the bound region (region *I*) up to a critical moisture content (M_c) but then increases sharply for high moisture contents. Therefore, the effect of bound water on dielectric properties is negligible.

The interaction of food components with water is a significant factor in affecting their dielectric properties. The stronger the binding forces between protein or carbohydrates and water, the smaller the value of the dielectric constant and loss factor since free water in the system decreases. For this reason, adjusting the moisture content is the key factor in formulating microwaveable foods.

The increase in water increases the polarization, which increases both dielectric constant and loss factor. At low moisture contents, variation of dielectric properties with moisture content is small.

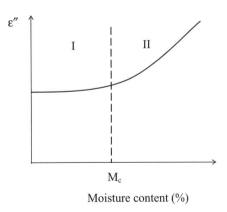

Figure 4.18   Variation of loss factor with moisture content.

There is a critical moisture content below which loss factor is not affected significantly with moisture content (Fig. 4.18). For food materials having high moisture contents, bound water does not play a significant role and the dielectric properties are affected by dissolved constituents as well as water content. In a recent study, it was shown that the dielectric loss factor of apples increased rapidly at water activity of around 0.9 which was explained by the contribution of greater amounts of mobile water to dielectric loss mechanisms (Martin-Esparza, Martinez-Navarrette, Chiralt, & Fito, 2005).

Dielectric properties of foods decrease during drying, since free moisture content in the system decreases. Feng, Tang, and Cavalieri (2002) showed that both dielectric constant and loss factor of apples decreased during drying owing to the reduced moisture content in the food.

4.3.4   Effects of Temperature on Dielectric Properties

Free and bound moisture content and ionic conductivity affect the rate of change of dielectric constant and loss factor with temperature. If the water is in bound form, the increase in temperature increases the dielectric properties. However, in the presence of free water, dielectric properties of free water decrease as temperature increases. Therefore, the rate of variation of dielectric properties depends on the ratio of bound to free moisture content.

During thawing, both dielectric constant and loss factor show large increases with temperature. After the material thaws, dielectric properties decrease with increasing temperature for different food materials except for a salted food (ham) (Fig. 4.19). The loss factor of ham shows a continuous increase during heating. The increase in loss factor with temperature was also observed in turkey meat, which contains high amounts of ash (Sipahioglu et al., 2003b).

The variation of dielectric loss factor of a salt solution or a salty material with respect to temperature is different because the loss factor of a salt solution is composed of two components: dipolar loss and ionic loss. Figure 4.20 shows the variation of loss factor components with temperature. Dipolar loss decreases with temperature at frequencies used in microwave processing. In contrast to dipolar loss, loss factor from ionic conduction increases with temperature owing to the decreased viscosity of the liquid and increased mobility of the ions. At higher temperatures, ions become more mobile and not tightly bound to water, and thus the loss factor from ionic loss component increases with temperature. On the other hand, microwave heating of water molecules or food containing free moisture decreases with increasing temperature (Prakash, 1991). The reasons for this are the rare hydrogen bonds and more intense movements which require less energy to overcome intermolecular bond at higher temperatures. For materials containing both dipolar and ionic components, it is possible to observe first a decrease and then an increase in loss factor with temperature.

There are limited dielectric properties data for foods below freezing temperatures. Data obtained for frozen foods and during melting of these foods are important to achieve uniform heating and prevent runaway heating in microwave thawing and tempering. Sipahioglu et al. (2003a) investigated the effects of moisture and ash content on dielectric properties of ham below and above freezing temperatures (-35 to $70°C$). Frozen ham samples had low dielectric properties until melting started at $-20°$ to $-10°C$. After melting took place, loss factor of ham increased with ash content (Sipahioglu et al., 2003a). On the contrary, increasing ash content reduced dielectric constant of a ham sample (Sipahioglu & Barringer, 2003). This can be explained by the fact that salts are capable of binding water which decreases the amount of water available for polarization. Ash content was not found

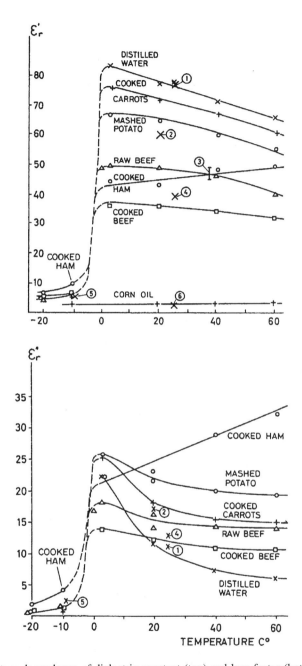

Figure 4.19 Temperature dependence of dielectric constant (top) and loss factor (bottom) of selected foods. [From Bengtsson, N.E., & Risman, P.O. Dielectric properties of foods at 3 GHz as determined by cavity perturbation technique. II Measurements in food materials. *Journal of Microwave Power, 6,* 107–123. Copyright © (1971) with permission of International Microwave Power Institute.]

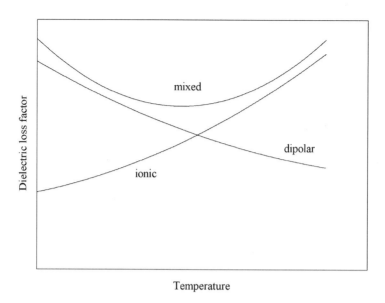

Figure 4.20   Variation of loss factor components with temperature.

---

to be significant in affecting dielectric constant of fruits since ash concentration was low in fruits (Sipahioglu & Barringer, 2003).

Dielectric properties data above the boiling point of water are also limited in the literature. Dielectric properties at high temperatures are important for microwave sterilization and pasteurization. Dielectric properties of whey protein gel, liquid whey protein mixture, macaroni noodles, and macaroni and cheese mixture were measured over a temperature range of 20 to 121°C at different frequencies (Wang et al., 2003). The dielectric constant of all samples except noodles decreased as temperature increased at frequencies of 915 and 1800 MHz. The increase in dielectric constant of cooked macaroni noodles with temperature is due to its low moisture content. There was a mild increase in the loss factor of samples with temperature.

4.3.5   Effects of Composition of Foods on Dielectric Properties

Dielectric properties of food products depend on composition. Carbohydrate, fat, moisture, protein, and salt contents are the major food components. The presence of free and bound water, surface charges, electrolytes, nonelectrolytes, and hydrogen bonding in the food product affect the dielectric properties. The physical changes that take place during processing such as moisture loss and protein denaturation also have an effect on dielectric properties. Therefore, the investigation of dielectric behavior of major food components and the effects of processing on dielectric properties are important for food technologists and engineers to improve the quality of microwave foods, to design microwaveable foods, and to develop new microwave processes.

Food components such as proteins, triglycerides, and starches have low dielectric activities at microwave frequencies. On the other hand, free water, monosaccharides, and ions have high dielectric activity (Shukla & Anantheswaran, 2001).

4.3.5.1 *Dielectric Properties of Salt Solutions*

Salt is one of the major components in food systems, which is responsible for ionic conduction. As can be seen in Fig. 4.21, addition of salt to sturgeon caviar decreased dielectric constant but increased loss factor. The decrease in dielectric constant with the addition of salt is due to binding of water in the system which reduces the available water for polarization. On the other hand, addition of salt increases the loss factor since more charged particles are added to the system and charge migration is increased. Both dielectric constant and loss factor increased with temperature but this increase was steeper for sturgeon caviar to which salt had been added (Al-Holy, Wang, Tang, & Raco, 2005). Nelson and Datta (2001) showed that the loss factor of salt solutions may increase or decrease with increasing temperature for different salt concentrations. As discussed previously, the loss factor of an aqueous ionic solution is represented by the addition of two components, dipole loss component and ionic loss component (Mudgett, 1995). The dipole loss component decreases but the ionic loss component increases with increasing temperature. The variation of loss factor with temperature depends on which mechanism is dominant. It was shown that for salt concentrations less than 1.0%, the loss factor decreased with temperature (Nelson & Datta, 2001). On the other hand, the loss factor increased as temperature increased for higher salt concentrations since ionic loss is the dominant mechanism.

Since the dissociated or ionized form of electrolytes interact with microwaves, the effects of pH and ionic strength become significant. Significant hydrogen ions can determine the degree of ionization. Dissociated ions migrate when exposed to an electric field and hence pH becomes an important factor in microwave heating. Ionic strength determines collision frequency. The collision increases at high concentrations of ions to a point that loss factor may have a positive temperature coefficient. Therefore, ionizable materials that increase surface temperature of the product are used in browning formulations for microwaveable products (Shukla & Anantheswaran, 2001).

4.3.5.2 *Dielectric Properties of Carbohydrates*

Starches, sugars, and gums are the major carbohydrates in food systems. For carbohydrate solutions, the effect of free water on dielectric properties becomes significant since carbohydrates themselves have small dielectric activities at microwave frequencies. Hydrogen bonds and hydroxyl–water interactions also play a significant role in dielectric properties of high sugar, maltodextrin, starch hydrolysate, and lactose such as disaccharide-based foods (Roebuck, Goldblith, & Westphal, 1972).

(a) Starch. Variation of dielectric properties of starch with temperature depends on whether starch is in solid state or in suspension form. Various researchers have studied the dielectric properties of starch in solid state and/or in suspension form (Moteleb, 1994; Ndife, Sumnu, & Bayindirli, 1998; Piyasena et al., 2003; Roebuck et al., 1972; Ryynänen, Risman, & Ohlsson, 1996). When the dielectric properties of different starches in powder form were measured at 2450 MHz, both the dielectric constant and the loss factor increased with temperature (Ndife et al., 1998). The difference between loss factors of different starches in powder form was explained by the differences in their bulk densities (Ndife et al., 1998). Table 4.2 shows the variation of loss factor of starch with bulk densities. The lower the bulk density, the lower the loss factor observed. Other researchers also found loss factors of other granular materials to be dependent on bulk density (Calay et al., 1995; Nelson, 1983).

For starch suspensions, the effect of free water on dielectric properties becomes significant. Dielectric constant and loss factor of different starch suspensions were shown to decrease as temperature and starch concentration increased (Ndife et al., 1998; Ryynänen et al., 1996). The dielectric properties

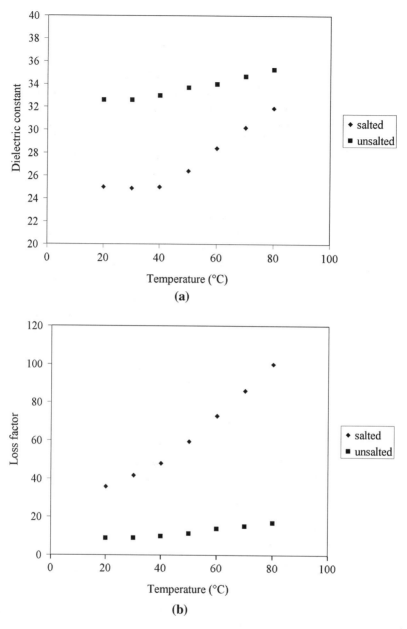

Figure 4.21 Dielectric properties. (a) Dielectric constant; (b) loss factor for salted (3%) and unsalted (0.2%) sturgeon caviar as a function of temperature at 915 MHz. [From Al-Holy, M., Wang, Y., Tang, J., & Raco, B. Dielectric properties of salmon (*Onecorhynchus keta*) and sturgeon (A*cipenser transmontanus*) caviar at radiofrequency (RF) and microwave (MW) pasteurization frequencies. *Journal of Food Engineering, 70,* 564–570. Copyright © (2005) with permission of Elsevier.]

Table 4.2 Bulk Density and Dielectric Properties of Dry Granular Starches at 30°C[a]

Starch Type	Bulk Density (g/cm³)	Dielectric Constant	Loss Factor
Corn	0.810	2.74	0.14
Rice	0.678	1.25	0.00
Tapioca	0.808	2.25	0.08
Wheat	0.790	2.42	0.05
Waxymaize	0.902	2.81	0.43
Amylomaize	0.886	2.42	0.37

[a]From Ndife, M., Sumnu, G., & Bayindirli, L. Dielectric properties of six different species of starch at 2450 MHz. *Food Research International, 31*, 43–52. Copyright © (1998) with permission of Elsevier.

of aqueous solutions are inversely related with temperature in the absence of ions, as discussed previously. The increase in starch concentration decreases both the dielectric constant and loss factor since starch molecules bind water and reduce the amount of free water in the system. The dielectric loss factors of different starch suspensions were also found to be a function of starch type (Ndife et al., 1998). Wheat, rice, and corn starches had significantly higher loss factor than tapioca, waxymaize, and amylomaize starches (Fig. 4.22) which may be related to the moisture binding properties of these starches. It is advisable to use starches having high dielectric properties in microwave baked products where poor starch gelatinization resulting from short baking time needs to be avoided. High dielectric properties of starch should be accompanied with the low thermal properties such as gelatinization enthalpy and specific heat capacity to achieve sufficient gelatinization in the product during baking.

The effects of temperature, concentration, frequency, and salt addition on dielectric properties of starch solutions have recently been studied and correlations have been developed for estimation of dielectric properties of starch solutions (Piyasena et al., 2003). Dielectric loss factor increased with increasing temperature and salt concentration. Dielectric constant of starch solutions containing no salt decreased with temperature. Salt ions affected the dielectric properties, especially the dielectric loss factor significantly. The dielectric constants of the salt solutions are known to decrease whereas the dielectric loss factor is known to increase with an increase in salt concentration.

Gelatinization of starch is an important physical phenomenon that affects dielectric properties. When the dielectric properties of gelatinized and ungelatinized potato starch were compared, dielectric constant of gelatinized potato starch was found to be higher than that of ungelatinized starch (Roebuck et al., 1972). Gelatinized starch binds less water to its structure which makes more free water to respond to the alternating electric field. This explains the higher dielectric properties of gelatinized potato as compared to the ungelatinized one.

(b) Sugar. Sugar is an important microwave absorbing food ingredient as compared to other hydrocolloids. Sugars modify the dielectric behavior of water. The hydroxyl water interactions stabilize liquid water by hydrogen bonds and affect the dielectric properties of sugar solutions. The degree of

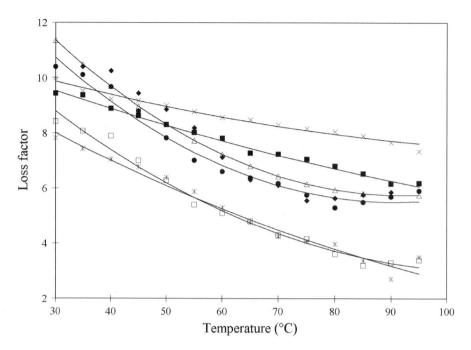

Figure 4.22   Variation of loss factor of starches with temperature for starch water ratio of 1:2 (□): waxymaize[d*], (*): amylomaize[d], (■): corn[b], (△): wheat[b], (●): tapioca[c], (x): rice.[a] *Starches with different letters have significantly different loss factor values. [From Ndife, M., Sumnu, G., & Bayindirli, L. Dielectric properties of six different species of starch at 2450 MHz. *Food Research International, 31*, 43–52. Copyright © (1998) with permission of Elsevier.]

microwave interaction depends on the extent of hydrogen bonding. Hydroxyl groups of glucose are more accessible for hydrogen bonding as compared to starches. In starches, less hydroxyl groups are exposed to water and fewer stable hydrogen bonds are formed. Therefore, the loss factors of the starch solutions were reported to be lower than that of the sugar solutions (Roebuck et al., 1972).

Dielectric properties of sugar solutions have been studied by various researchers (Liao et al., 2001, 2003; Roebuck et al., 1972). Dielectric properties of glucose solutions having different concentrations (10–60%) were found to be a function of temperature and composition (Liao et al., 2003). Dielectric constant of glucose solution increased but the loss factor of glucose solution decreased with temperature. Increasing glucose concentration decreased dielectric constant since less water was free to respond the electric field. There is a critical sugar concentration that affects the dielectric loss factor of sugar solution. When the temperatures exceeded 40°C, loss factor increased with an increase in concentration since more hydrogen bonds are stabilized by the presence of more hydroxyl groups of sugars. However, at lower temperatures glucose solution became saturated at lower concentration and loss factor decreased with concentration.

(c)   Gum. Gums have the ability to bind high amounts of free water in the system. Therefore, depending on the amount of moisture bound to the gums, dielectric constant and loss factor of the system change.

Charge of the gum is a significant factor in affecting its dielectric properties. As the charge increases, the amount of moisture bound to the charged groups increases, which lowers the dielectric constant and loss factor (Prakash, Nelson, Mangino, & Hansen, 1992). In the absence of water the effect of charge disappears. The effect of charge on dielectric values may be due to the fact that water associated with highly hydrophilic charged groups may not be free to interact with microwaves.

For microwaveable food formulations, it is important to know water binding capacity of the gums and viscosity of the solution to have an idea about the dielectric properties and microwave heatability of these formulations. When hydrocolloids are used in the range of 0.1% to 2.0%, they can immobilize 25% to 60% of water (Shukla & Anantheswaran, 2001). Since hydrocolloids can bind different amounts of water, food formulations containing one or more than one hydrocolloid are expected to have different amounts of free water in the system, which can affect polarization. Therefore, interaction of food with microwaves is expected to change in the presence of gums.

4.3.5.3 Dielectric Properties of Proteins

Free amino acids are dielectrically reactive (Pething, 1979). Free amino acids and polypeptides contribute to the increase in dielectric loss factor. Since protein dipole moments are a function of their amino acids and pH of the medium, the dielectric properties and microwave reactivity of cereal, legume, milk, meat, and fish proteins are expected to be different. The water adsorbed on the protein also affects their dielectric properties (Shukla & Anantheswaran, 2001).

Dielectric properties of proteins change during denaturation. Protein denaturation is defined as the physical change of the protein molecule due to heat, UV, or agitation which results in a reduction in protein solubility and increase in solution viscosity (McWilliams, 1989). During denaturation of proteins, since the structure of protein is disturbed, the asymmetry of the charge distribution will increase. This will result in large dipole moment and polarization, which will affect the dielectric properties. Moreover, moisture is either bound by the protein molecule or released to the system during denaturation which shows a decrease or increase in dielectric properties, respectively. There are various studies showing that the dielectric properties can be used to understand protein denaturation (Bircan & Barringer, 2002a,b; Bircan et al., 2001). In these studies, dielectric properties were found to be as effective as DSC for determination of denaturation temperature. The comparison of determination of denaturation temperature by dielectric properties and by DSC is shown in Table 4.3.

The loss factor of egg yolk protein increased and then decreased with temperature by making a peak during denaturation (Bircan & Barringer, 2002a). The increase in loss factor with temperature may be due to the presence of ions that egg yolk contains. The reduction of loss factor after denaturation is due to the binding of water and decrease in mobility of ions. On the contrary, the loss factor of meat protein actomyosin increased during protein denaturation owing to the release of water during denaturation (Bircan & Barringer 2002b). Dielectric properties of gluten protein were also shown to be affected by heating (Umbach, Davis, Gordon, & Callaghan, 1992). The dielectric constant and loss factor of heated gluten–starch mixture were found to be less than unheated mixture. As the amount of gluten protein in the system increased, there was a decrease in the dielectric constant but the loss factor remained constant. The interaction of gluten with microwaves has been known to adversely affect the texture of microwave baked breads (Yin & Walker, 1995). Microwave-baked breads containing low amounts of gluten were shown to be softer than the ones containing high amounts of gluten (Ozmutlu, Sumnu, & Sahin, 2001). Addition of low levels of gliadin, mildly hydrolyzed wheat gluten, or wheat protein isolate to the bread formula was effective in reducing the microwave induced toughness of pup loaf bread but was not effective in reducing microwave induced toughness of hoagie buns (Miller, Maningat, & Bassi, 2003).

Table 4.3 The Denaturation Temperatures Determined by Dielectric Properties and by DSC[a]

Sample	Denaturation (°C)	
	Dielectrics	*DSC*
20% Whey solution	75–80	78.6
20% Whey + 5% sugar	75–80	79.3
20% Whey + 15% sugar	80–85	82.4
20% Whey + 2% salt	83.8	81.2
20% Whey at pH 4	85–90	85.5
10% β-Lactoglobulin	75–80	78.8
20% α-Lactalbumin	70–75	75.0
10% Bovine serum albumin	85–90	87.6

[a]From Bircan, C., Barringer, S.A., & Mangino, M.E. Use of dielectric properties to detect whey protein denaturation. *Journal of Microwave Power and Electromagnetic Energy, 36*, 179–186. Copyright © (2001) with permission of International Microwave Power Institute.

4.3.5.4   Dielectric Properties of Fat

Since lipids are hydrophobic except for ionizable carboxyl groups of fatty acids, they do not interact much with microwaves (Mudgett & Westphal, 1989). Therefore, dielectric properties of fats and oils are very low (Figs. 4.16 and 4.17). The effect of fat on dielectric properties of food systems is mainly the result of their dilution effect in the system. The increase in fat content reduces the free water content in the system, which reduces the dielectric properties (Ryynänen, 1995).

4.3.6   Assessment of Quality of Foods by Using Dielectric Properties

Dielectric properties can be used for quality control of foods such as to determine the state of fish and meat freshness, to evaluate frying oil quality, and to determine moisture content.

Dielectric properties of meat are expected to change during rigor mortis. Rigor mortis is defined as the temporary rigidity of muscles that develops after death of an animal (McWilliams, 1989). A postmortem decrease in pH brings proteins closer to their isoelectric point and causes a decrease in water holding capacity (Parasi, Franci, & Poli, 2002), which is important for polarization.

It was found that haddock muscle showed significant changes in dielectric properties during rigor mortis (Martinsen, Grimnes, & Mirtaheri, 2000). Parasi et al. (2002) concluded that the state of freshness of sea bass can be evaluated by the measurement of dielectric properties.

Dielectric properties can also be used for evaluation of frying oil quality (Fritsch, Egberg, & Magnusun, 1979; Inoue, Hagura, Ishikawa, & Suzuki, 2002; Shi, Lung, & Sun, 1998). Hein, Henning, and Isengard (1998) used dielectric properties to determine heat abuse for frying oils and fats. Suitability of dielectric constant for continuous evaluation of oil quality was investigated and the relationship with dielectric properties and other evaluation parameters such as acid value, density, and viscosity were determined (Inoue et al., 2002). Dielectric constant was found to be in good correlation with acid value, density, and relative viscosity. Dielectric constant of soybean oil was proposed to be considered for a new degradation parameter comparable with acid value.

Since the dielectric properties of foods are dependent on moisture content, they can be used to measure the moisture content in grains and seeds (Chen & Sun, 1991). The dielectric properties give a rapid, nondestructive sensing of moisture content of agricultural products (Kuang & Nelson, 1997). Moisture of agricultural products could be determined indirectly by measuring dielectric properties by using the relationship of dielectric constant with moisture content, frequency, and bulk density. The feasibility of a microwave sensor for water activity measurements was investigated by Clerjon, Daudin, and Damez, (2003). As a result, a correlation was determined between water activity and dielectric properties of animal gelatin gel, which needs further research on real food samples. Raweendranath and Mathew (1995) suggested that dielectric behavior of water can be an effective method for detection of pollutants in water at microwave frequency of 2.685 GHz.

4.3.7 Measurement of Dielectric Properties

Methods for measurement of dielectric properties of food materials depend on the food being tested, degree of accuracy and frequency. Dielectric property measurement methods can be categorized as reflection or transmission types depending on resonant or nonresonant systems with closed or open structures (Kraszewski, 1980). Dielectric measurement of a sample is simple (Engelder & Buffler, 1991):

1. A microwave signal is generated at the frequency of interest.
2. Signal is directed through the sample.
3. Changes in the signal caused by the sample are measured.
4. From these changes the dielectric constant and loss factor are determined.

The most commonly used methods for dielectric measurement of foods are the transmission line method, coaxial probe method, cavity perturbation method, and free space transmission method.

Figure 4.23 is a schematic diagram of the transmission line method. In this method, dielectric constant and loss factor are derived from transmission line theory. These parameters are determined from the phase and amplitude of a microwave signal from a sample placed against the end of a short-circuited transmission line such as a waveguide or a coaxial line. When a microwave signal is sent

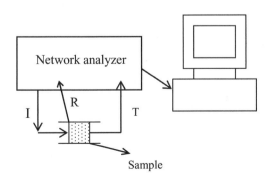

Figure 4.23 Schematic diagram of the transmission line method.

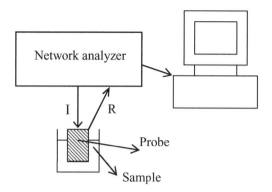

Figure 4.24 Schematic diagram of the coaxial probe method.

from the network analyzer, some of the signal is reflected from the sample; the rest is absorbed by the sample or transmitted through it. For a waveguide structure, rectangular samples that fit into the dimensions of waveguide at the frequency being measured are necessary. For coaxial lines, an annular sample should be used. Coaxial lines can be used for a wide range of microwave frequencies but waveguides have narrow ranges and as the frequency changes different sample sizes should be used.

The coaxial probe method is a modification of the transmission line method. This method is based on an open-ended coaxial line inserted into sample to be measured (Fig. 4.24). It can be used for a wide frequency range. This method is nondestructive and ideal for liquids and semisolids. On the other hand, the probe is not recommended to be used for samples having low dielectric constant and loss factor such as fats and oils. It does not require a definite sample shape. A typical system for dielectric properties measurement includes a probe, a network analyzer, a coaxial cable, PC, and software. Agilent 85070E dielectric kit is a commercialized system of this method. To use this kit, the minimum recommended loss tangent, which is the ratio of loss factor to dielectric constant, should be greater than 0.05. There are two types of probes used in coaxial probe method: high-temperature and slim form probe. Operating temperatures are −40 to 200°C and 0 to 125°C for high-temperature and slim form probes, respectively. The required thickness of the sample is critical. The thickness of the sample should be greater than $\left| \dfrac{20}{\sqrt{\varepsilon_r^*}} \right|$, to use the high-temperature probe where ε_r^* is relative complex permittivity. To use the slim form probe, there should be a 5-mm insertion and 5 mm around the probe tip.

Cavity perturbation method can be very accurate for measuring low values of loss factor such as fats and oils. This method has the disadvantage of providing results at only one frequency. The measurement is made by placing the sample through the center of a waveguide that has been made into a cavity (Fig. 4.25). A recommended waveguide cavity design is available as a standard procedure given by American Society for Testing and Materials (ASTM, 1986). Sample geometries can be circular, rectangular, or square cross section with a recommended dimension less than 0.318 cm for a frequency of 2450 MHz. Changes in center frequency and width due to the insertion of the sample provide information to calculate dielectric constant. The changes in ratio of energy stored to the energy dissipated are used to determine loss factor.

Free space transmission method has the advantage of being a noninvasive and noncontact measuring method. Both transmission line and cavity methods can be implemented in the free space. A sample is placed between a transmitting antenna and a receiving antenna (Fig. 4.26). The attenuation and phase

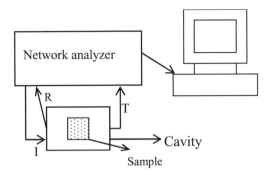

Figure 4.25 Schematic diagram of the cavity perturbation method.

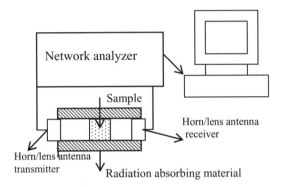

Figure 4.26 Schematic diagram of the free space transmission method.

shift of the signal are measured and dielectric constant and loss factor are calculated from these results. A vector network analyzer is usually required to measure reflection and transmission coefficients. A software program is used to convert measured data to dielectric constant and loss factor. In this method, no sample preparation is necessary. Therefore, it can be used for continuous monitoring and online control (Kraszewski, Trabelsi, & Nelson, 1995). More details about this method can be found in Trabelsi and Nelson (2003).

REFERENCES

Al-Holy, M., Wang, Y., Tang, J., & Raco, B. (2005). Dielectric properties of salmon (*Onecorhynchus keta*) and sturgeon (A*cipenser transmontanus*) caviar at radiofrequency (RF) and microwave (MW) pasteurization frequencies. *Journal of Food Engineering, 70*, 564–570.

Altunakar, B., Sahin, S., & Sumnu, G. (2004). Functionality of batters containing different starch types for deep-fat frying of chicken nuggets. *European Food Research and Technology, 218*, 317–322.

ASTM (1986). Standard methods of test for complex permittivity (dielectric constant) of solid electrical insulating materials at microwave frequencies and temperatures to 1650°C. Document D2520-86 (reapproved 1990). Philadelphia: American Society for Testing and Materials.

Baik, O.D., & Mittal, G.S. (2003). Kinetics of tofu color changes during deep-fat frying. *Lebensmittel Wissenschaft und Technologie, 36*, 43–48.

Baysal, T., Icier, F., Ersus, S., & Yildiz, H. (2003). Effects of microwave and infrared drying on the quality of carrot and garlic. *European Food Research and Technology, 218*, 68–73.

Bengtsson, N.E., & Risman, P.O. (1971). Dielectric properties of foods at 3 GHz as determined by cavity perturbation technique. II. Measurements in food materials. *Journal of Microwave Power, 6*, 107–123.

Billmeyer, F.W., & Saltzman, M. (1981). *Principles of Color Technology*, 2nd ed. New York: John Wiley & Sons.

Bircan, C., & Barringer, S.A. (2002a). Use of dielectric properties to detect whey protein denaturation. *Journal of Microwave Power and Electromagnetic Energy, 37*, 89–96.

Bircan, C., & Barringer, S.A. (2002b). Determination of protein denaturation of muscle foods using dielectric properties. *Journal of Food Science, 67*, 202–205.

Bircan, C., Barringer, S.A., & Mangino, M.E. (2001). Use of dielectric properties to detect whey protein denaturation. *Journal of Microwave Power and Electromagnetic Energy, 36*, 179–186.

Calay, R.K., Newborough, M., Probert, D., & Calay, P.S. (1995). Predictive equations for dielectric properties of foods. *International Journal of Food Science and Technology, 29*, 699–713.

Chen, P., & Sun, Z. (1991). A review of nondestructive methods for quality evaluation and sorting of agricultural products. *Journal of Agricultural Engineering Research, 49* (2), 85–98.

Choi, L.H., & Nielsen, S.S. (2005). The effect of thermal and nonthermal processing methods on apple cider quality and consumer acceptability. *Journal of Food Quality, 28*, 13–29.

Clerjon, S., Daudin, J.D., & Damez, J.L. (2003). Water activity and dielectric properties of gels in the frequency range 200 MHz-6 GHz. *Food Chemistry, 82*, 87–97.

Datta, A.K., Sumnu, G., & Raghavan, G.S.V. (2005). Dielectric properties of foods. In M.A. Rao, S.S.H. Rizvi & A.K. Datta (Eds.), *Engineering Properties of Foods,* 3rd ed. (pp. 501–565). Boca Raton, FL: CRC Press Taylor & Francis.

Engelder, S.D., & Buffler, R.C. (1991). Measuring dielectric properties of food products at microwave frequencies. *Microwave World, 12*, 6–14.

Feng, H., Tang, J., & Cavalieri, R.P. (2002). Dielectric properties of dehydrated apples as affected by moisture and temperature. *Transactions of the ASAE, 45*, 129–135.

Francis, F.J. (1962). Relationship between flesh color and pigment content in squash. *Proceedings of the American Society of Horticultural Science, 81*, 408–414.

Francis, F.J. (1983). Colorimetry of foods. In M. Peleg, & E.B. Bagley, (Eds), *Physical Properties of Foods* (pp. 105–123). Westport, CT: AVI.

Francis, F.J. (2005). Colorimetric properties of foods. In M.A. Rao, S.S.H. Rizvi, & A.K. Datta (Eds.), *Engineering Properties of Foods,* 3rd ed. (pp. 703–732). Boca Raton: CRC Press Taylor & Francis.

Fritsch, C.W., Egberg, D.C., & Magnusun, J.S. (1979). Changes in dielectric constant as a measure of frying oil deterioration. *Journal of the American Oil Chemists Society, 56*, 746–750.

Guan, D., Cheng, M., Wang, Y., & Tang, J. (2004). Dielectric properties of mashed potatoes relevant to microwave and radio-frequency pasteurization and sterilization processes. *Journal of Food Science, 69*, E30–E37.

Hein, M., Henning, H., & Isengard H.D. (1998). Determination of total polar parts with new methods for the quality survey of frying fats and oils. *Talanta, 47*, 447–454.

Hu, X.P., & Mallikarjunan, P. (2005). Thermal and dielectric properties of shucked oysters. *Lebensmittel Wissenschaft und Technologie, 38*, 489–494.

Inoue, C., Hagura, Y., Ishikawa, M., & Suzuki K. (2002). The dielectric property of soybean oil in deep-fat frying and effect of frequency. *Journal of Food Science, 67*, 1126–1129.

Keskin, S.O., Sumnu, G., & Sahin, S. (2004). Bread baking in halogen lamp-microwave combination oven. *Food Research International, 37*, 489–495.

Kraszewski, A.W. (1980). Microwave aquametry—a review. *Journal of Microwave Power, 15*, 209–220.

Kraszewski, A.W., Trabelsi, S., & Nelson, S.O. (1995). Microwave dielectric properties of wheat. Proceedings of 30th Microwave Power Symposium, July 9–12, Denver CO, pp. 90–93.

Kuang, W., & Nelson, S.O. (1997). Dielectric relaxation characteristic of fresh fruits and vegetables from 3 to 20 GHz. *Journal of Microwave Power & Electromagnetic Energy, 32*, 114–122.

Kumar, D.G.P., Hebbar, H.U., Sukumar, D., & Ramesh, M.N. (2005). Infrared and hot-air drying of onions. *Journal of Food Processing and Preservation, 29*, 132–150.

Liao, X., Raghavan, V.G.S., Meda, V., & Yaylayan, V.A. (2001). Dielectric properties of supersaturated α-D glucose aqueous solutions at 2450 MHz. *Journal of Microwave Power and Electromagnetic Energy, 36*, 131–138.

Liao, X.J., Raghavan, G.S.V., Dai, J., & Yaylayan, V.A. (2003). Dielectric properties of α-D glucose aqueous solutions at 2450 MHz. *Food Research International, 36*, 485–490.

Lyng, J.G., Zhang, L., & Brunton, N.P. (2005). A survey of the dielectric properties of meats and ingredients used in meat product manufacture. *Meat Science, 69*, 589–602.

MacKinney, G., & Little, A.C. (1962). *Color of Foods*. Westport, CT: AVI.

Marcus, R.T. (1998). The measurement of color. In K. Naussau (Ed.), *Color for Science, Art and Technology* (pp. 34–96). Amsterdam: Elsevier Science.

Martin-Esparza, M.E., Martinez-Navarrette, N., Chiralt, A., & Fito, P. (2005). Dielectric behavior of apple (var. Granny Smith) at different moisture contents: Effect of vacuum impregnation. *Journal of Food Engineering* (doi:10.1016/j.jfoodeng.2005.06.018)

Martinsen, O.G., Grimnes, S., & Mirtaheri, P. (2000). Non-invasive measurements of post-mortem changes in dielectric properties of haddock muscle-A pilot study. *Journal of Food Engineering, 43*, 189–192.

Maskan, M. (2001). Kinetics of colour change of kiwifruits during hot air and microwave drying. *Journal of Food Engineering, 48*, 169–175.

McLaren, K. (1976). The development of C.I.E. (L*, a*, b*) uniform color space and color-difference formula. *Journal of the Society of Dyers and Colourists, 92*, 338–341.

McWilliams, M. (1989). *Foods: Experimental Perspectives*. New York: Macmillan.

Miller, R.A., Maningat, C.C., & Bassi, S.D. (2003). Effect of gluten fractions in reducing microwave-induced toughness of bread and buns. *Cereal Foods World, 48*, 76–77.

Moteleb, M.M.A. (1994). Some of the dielectric properties of starch. *Polymer International, 35*, 243–247.

Moyono, P.C., Rioseco, V.K., & Gonzalez, P.A. (2002). Kinetics of crust color changes during deep-fat frying of impregnated french fries. *Journal of Food Engineering, 54*, 249–255.

Mudgett. R.E. (1995). Electrical Properties of Foods. In M.A. Rao & S.S.H. Rizvi (Eds.), *Engineering Properties of Foods*, 2nd ed. (pp. 389–455). New York: Marcel Dekker.

Mudgett, R.E., & Westphal, W.B. (1989). Dielectric behavior of an aqueous cation exchanger. *Journal of Microwave Power & Electromagnetic Energy, 24*, 33–37.

Ndife, M., Sumnu, G., & Bayindirli, L. (1998). Dielectric properties of six different species of starch at 2450 MHz. *Food Research International, 31*, 43–52.

Nelson, S. (1983). Density dependence of dielectric properties of particulate materials. *Transactions of the ASAE, 26*, 1823–1825, 1829.

Nelson, S.O., & Datta, A.K. (2001). Dielectric properties of food materials and electric field interactions. In A.K. Datta & R.C. Anantheswaran (Eds.), *Handbook of Microwave Technology for Food Applications* (pp. 69–114). New York: Marcel Dekker.

Ongen, G., Sargin, S., Tetik, D., & Kose, T. (2005). Hot air drying of green table olives. *Food Technology and Biotechnology, 43*, 181–187.

Ozmutlu, O., Sumnu, G., & Sahin, S. (2001). Effects of different formulations on the quality of microwave baked bread. *European Food Research and Technology, 213*, 38–42.

Parasi, G., Franci, O., & Poli, B.M. (2002). Application of multivariable analysis to sensorial and instrumental parameters of freshness in refrigerated sea bass (*Dicentrarchus labrax*) during shelf life. *Aquaculture, 214*, 153–167.

Pething, R. (1979). *Dielectric and Electronic Properties of Biological Materials*. New York: John Wiley & Sons.

Piyasena, P., Ramaswamy, H.S., Awuah, G.B., & Defelice, C. (2003). Dielectric properties of starch solutions as influenced by temperature, concentration, frequency and salt. *Journal of Food Process Engineering, 26*, 93–119.

Prakash, A. (1991). *The Effect of Microwave Energy on the Structure and Function of Food Hydrocolloids*. M.S thesis, The Ohio State University.

Prakash, A., Nelson, S.O., Mangino, M.E., & Hansen, P.M.T. (1992). Variation of microwave dielectric properties of hydrocolloids with moisture content, temperature and stoichiometric charge. *Food Hydrocolloids, 6*, 315–322.

Ramirez-Jimenez, A., Garcia-Villanova, B., & Guerra-Hernandez, E. (2001). Effect of toasting time on the browning of sliced bread. *Journal of the Science of Food and Agriculture, 81*, 513–518.

Rattanathanalerk, M., Chiewchan, N., & Srichumpoung, W. (2005). Effect of thermal processing on the quality loss of pineapple juice. *Journal of Food Engineering, 66*, 259–265.

Raweendranath, U., & Mathew, K.T. (1995). Microwave technique for water pollution study. *Journal of Microwave Power & Electromagnetic Energy, 30*(3), 188–194.

Regier, M., Housova, J., & Hoke, K. (2001). Dielectric properties of mashed potatoes. *International Journal of Food Properties, 4*, 431–439.

Roebuck, B.D., Goldblith, S.A., & Westphal, W.B. (1972). Dielectric properties of carbohydrate water mixtures at microwave frequencies. *Journal of Food Science, 37*, 199–204.

Ryynänen, S. (1995). The electromagnetic properties of food materials: A review of basic principles. *Journal of Food Engineering, 26*, 409–429.

Ryynänen, S., Risman, P.O., & Ohlsson, T. (1996). The dielectric properties of Native starch solutions-A research note. *Journal of Microwave Power & Electromagnetic Energy, 31*, 50–53.

Sevimli, M., Sumnu, G., & Sahin, S. (2005). Optimization of halogen lamp-microwave combination baking of cakes: A response surface study. *European Food Research and Technology, 221*, 61–68.

Shi, L.S., Lung, B.H., & Sun, H.L. (1998). Effects of vacuum frying on the oxidative stability of oils. *Journal of American Oil Chemists Society, 75*, 1393–1398.

Shukla, T P, & Anantheswaran, R C. (2001). Ingredient interactions and product development. In A.K. Datta, & R.C. Anantheswaran (Eds.), *Handbook of Microwave Technology for Food Applications* (pp. 355–395). New York: Marcel Dekker.

Sipahioglu, O., & Barringer, S.A. (2003). Dielectric properties of vegetables and fruits as a function of temperature, ash and moisture content. *Journal of Food Science, 68*, 234–239.

Sipahioglu, O., Barringer, S.A., & Taub, I. (2003a). Modelling the dielectric properties of ham as a function of temperature and composition. *Journal of Food Science, 68*, 904–908.

Sipahioglu, O., Barringer, S.A., Taub, I., & Yang, A.P.P. (2003b). Characterization and modeling of dielectric properties of turkey meat as a function of temperature and composition. *Journal of Food Science, 68*, 521–527.

Trabelsi, S., & Nelson, S.O. (2003). Free space measurement of dielectric properties of cereal grain and oil seed at microwave frequencies. *Measurement Science and Technology, 14*, 589–600.

Umbach, S.L., Davis, E.A., Gordon, J., & Callaghan, P.T. (1992). Water self-diffusion coefficients and dielectric properties determined for starch-gluten-water mixtures heated by microwave and conventional methods. *Cereal Chemistry, 69*, 637–642.

Vélez-Ruiz J.F., & Sosa-Morales, M.E. (2003). Evaluation of physical properties of dough of donuts during deep-fat frying at different temperatures. *International Journal of Food Properties, 6*, 341–353.

Venkatesh, M.S., & Raghavan, G.S.V. (2004). An overview of microwave processing and dielectric properties of agri-food materials. *Biosystems Engineering, 88*, 1–18.

Wang, Y., Wig, T.D., Tang, J., & Hallberg, L.M. (2003). Dielectric properties of foods relevant to RF and microwave pasteurization and sterilization. *Journal of Food Engineering, 57*, 257–268.

Yin, Y., & Walker, C.E. (1995). A quality comparison of breads baked by conventional versus non-conventional ovens: A review. *Journal of the Science of Food & Agriculture, 67*, 283–291.

Zhang, L., Lyng, J.G., Brunton, N., Morgan, D., & McKenna B. (2004). Dielectric and thermophysical properties of meat batters over a temperature range of 5–85° C. *Meat Science, 68*, 173–184.

CHAPTER 5

Water Activity and Sorption Properties of Foods

SUMMARY

Water activity and sorption properties of foods have been considered as important physical properties in food formulations and processes. Most of the biochemical and microbiological reactions are controlled by the water activity of the system, which is therefore a useful parameter to predict food stability and shelf life. The rate of moisture transfer in the drying process and through the packaging film or edible food coating during storage can be estimated and as a result drying conditions, packaging, or coating material can be selected using water activity and sorption properties of foods. In addition, these properties must be considered in product development.

There are many methods for measuring water activity of foods. This chapter provides information about the theory of water activity, its prediction and measurement methods, and preparation of sorption isotherms. Principles of colligative properties such as boiling point elevation, freezing point depression, and osmotic pressure are also discussed. In addition, the most commonly used sorption models are given.

Water activity of foods can be measured using methods based on colligative properties, isopiestic transfer, and hygroscopicity of salts and using hygrometers. Moisture sorption isotherm describes the relationship between water activity and the equilibrium moisture content of a food product at constant temperature. The simplest method to prepare sorption isotherms of foods is storing a weighed sample in an enclosed container maintained at a certain relative humidity at constant temperature and reweighing it after equilibrium is reached. The desired relative humidity environments can be generated by using saturated salt solutions, sulfuric acid, and glycerol.

5.1 CRITERIA OF EQUILIBRIUM

The criterion for thermal equilibrium is equality of temperatures while the criterion for mechanical equilibrium is equality of pressures. Physicochemical equilibrium is characterized by equality of chemical potential (μ) of each component. The chemical potential determines whether a substance will undergo a chemical reaction or diffuse from one part of the system to another.

The chemical potential of a component in the liquid (L) phase is equal to that in the vapor (V) phase if vapor and liquid are in equilibrium:

$$\mu_i^L = \mu_i^V \tag{5.1}$$

Chemical potential is the partial molar free energy and can be expressed as:

$$\mu_i = \left(\frac{\partial G}{\partial n_i}\right)_{T,P,nj} \tag{5.2}$$

where G is Gibbs free energy and n_i is the moles of component i. The definition shows that the chemical potential of a component of a homogenous mixture is equal to the ratio of the increase in Gibbs free energy on the addition of an infinitesimal amount of the substance.

Gibbs free energy is defined as a combination of enthalpy (H), temperature (T), and entropy (S):

$$G = H - TS \tag{5.3}$$

Enthalpy can be expressed in terms of internal energy (U), pressure (P), and volume (V):

$$H = U + PV \tag{5.4}$$

Inserting Eq. (5.4) into Eq. (5.3):

$$G = U + PV - TS \tag{5.5}$$

Differentiating Eq. (5.5):

$$dG = dU + PdV + VdP - TdS - SdT \tag{5.6}$$

For a reversible process in a closed system of constant composition, the first and second laws of thermodynamics may be combined to yield:

$$dU = TdS - PdV \tag{5.7}$$

Introducing Eq. (5.7) into (5.6):

$$dG = VdP - SdT \tag{5.8}$$

If the system is at constant composition, Eqs. (5.9) and (5.10) can be obtained from Eq. (5.8):

$$\left(\frac{\partial G}{\partial T}\right)_{P,ni} = -S \tag{5.9}$$

Since entropy of the system is always positive, Gibbs free energy decreases with increasing temperature at constant pressure.

$$\left(\frac{\partial G}{\partial P}\right)_{T,ni} = V \tag{5.10}$$

Since volume of a system is always positive, Gibbs free energy increases with increasing pressure at constant temperature.

For an open system in which there is an exchange of matter with its surroundings, total Gibbs free energy can be expressed as a function of number of moles of each chemical species present in the system in addition to temperature and pressure.

$$G = f(T, P, n_1, n_2, \ldots, n_i)$$

Total differential of Gibbs free energy is:

$$dG = \left(\frac{\partial G}{\partial T}\right)_{P,ni} dT + \left(\frac{\partial G}{\partial P}\right)_{T,ni} dP + \sum_{i=1}^{k} \left(\frac{\partial G}{\partial n_i}\right)_{T,P,nj} dn_i \tag{5.11}$$

where n_i means that the amounts of all substances are held constant and n_j means that the amounts of all substances are held constant except the one being varied ($j \neq i$).

Replacing the first and second partial derivatives of Eq. (5.11) by Eqs. (5.9) and (5.10), respectively:

$$dG = VdP - SdT + \sum_{i=1}^{k} \left(\frac{\partial G}{\partial n_i}\right)_{T,P,nj} dn_i \tag{5.12}$$

Remembering Eq. (5.2):

$$\mu_i = \left(\frac{\partial G}{\partial n_i}\right)_{T,P,nj} \tag{5.2}$$

At constant temperature and pressure, Eq. (5.12) is simplified as:

$$dG = \sum_{i=1}^{k} \mu_i dn_i \tag{5.13}$$

This equation shows that Gibbs free energy of a system is the sum of the contributions of various components.

The general criterion for a system to be at equilibrium is that:

$$(dG)_{T,P} = 0 \tag{5.14}$$

$$(dG)_{T,P} = \sum_{i=1}^{k} (\mu_i^L dn_i^L) + \sum_{i=1}^{k} (\mu_i^V dn_i^V) = 0 \tag{5.15}$$

If the system is closed:

$$dn_i^V = -dn_i^L \tag{5.16}$$

Then, Eq. (5.15) becomes:

$$\sum_{i=1}^{k} (\mu_i^L dn_i^L) + \sum_{i=1}^{k} (-\mu_i^V dn_i^L) = 0 \tag{5.17}$$

$$\Rightarrow \sum_{i=1}^{k} (\mu_i^L - \mu_i^V) dn_i^L = 0 \tag{5.18}$$

The only way that Eq. (5.18) can be satisfied is:

$$\mu_i^L = \mu_i^V \tag{5.1}$$

At the same temperature and pressure, the equilibrium condition is satisfied when the chemical potentials of each species in liquid and vapor phases are the same.

5.2   IDEAL SOLUTION—RAOULT'S LAW

A solution can be defined as ideal if the cohesive forces inside a solution are uniform. This means that in the presence of two components A and B, the forces between A and B, A and A, and B and B are all the same.

Equation (5.10) can be rewritten in terms of partial molar quantities and since the partial molar free energy is the chemical potential, for compound A in a solution:

$$\left(\frac{\partial \bar{G}}{\partial P_A}\right)_T = \bar{V}_A = \left(\frac{\partial \mu_A}{\partial P_A}\right)_T \tag{5.19}$$

where $\bar{V}_A$ is the molar volume of component A in solution which is the volume divided by the number of moles of A.

Using ideal gas law and Eq. (5.19), μ_A can be related to the partial vapor pressure by:

$$d\mu_A = \bar{V}_A dP_A = RT\frac{dP_A}{P_A} \tag{5.20}$$

If μ_A^0 is the value of chemical potential when the pressure is 1 atm, integrating Eq. (5.20):

$$\int_{\mu_A^0}^{\mu_A} d\mu_A = \int_1^{P_A} RT\frac{dP_A}{P_A} \tag{5.21}$$

$$\Rightarrow \mu_A = \mu_A^0 + RT\ln P_A \tag{5.22}$$

Partial vapor pressure of a component, which is a measure of tendency of the given component to escape from solution into the vapor phase, is an important property for solutions. For a solution in equilibrium with its vapor:

$$\mu_A^{so\ln} = \mu_A^{vapor} = \mu_A^0 + RT\ln P_A \tag{5.23}$$

Thus, chemical potential of component A in solution is related to the partial vapor pressure of A above the solution. Equation (5.23) is true only when vapor behaves as an ideal gas.

A solution is ideal if the escaping tendency of each component is proportional to the mole fraction of that component in the solution. The escaping tendency of component A from an ideal solution, as measured by its partial vapor pressure, is proportional to the vapor pressure of pure liquid A and mole fraction of A molecules in the solution. This can be expressed by Raoult's law as:

$$P_A = X_A P_A^0 \tag{5.24}$$

where P_A is the partial vapor pressure of A, X_A is its mole fraction, and P_A^0 is the vapor pressure of pure liquid A at the same temperature.

If component B is added to pure A, vapor pressure is decreased as:

$$P_A = (1 - X_B)P_A^0 \tag{5.25}$$

$$\Rightarrow X_B = \frac{P_A^0 - P_A}{P_A^0} \tag{5.26}$$

Equation (5.25) (relative pressure lowering) is useful for solutions of a relatively nonvolatile solute in a volatile solvent.

Inserting Eq. (5.24) into Eq. (5.23), Eq. (5.27) can be obtained:

$$\mu_A = \mu_A^0 + RT\ln P_A^0 + RT\ln X_A \tag{5.27}$$

5.3   HENRY'S LAW

Consider a solution containing solute B in solvent A. If the solution is very dilute, a condition is reached in which each molecule B is completely surrounded by component A. Solute B is then in a uniform environment irrespective of the fact that A and B may form solutions that are not ideal at higher concentrations. In such a case, the escaping tendency of B from its environment is proportional to its mole fraction, which can be expressed by Henry's law as:

$$P_B = kX_B \tag{5.28}$$

where k is the Henry's law constant.

Henry's law is not restricted to gas–liquid systems. It is valid for fairly and extremely dilute solutions.

5.4   COLLIGATIVE PROPERTIES

Colligative properties depend on the number of solute molecules or ions added to the solvent. Vapor pressure lowering, boiling point elevation, freezing point depression, and osmotic pressure are the colligative properties. These properties are used to determine the molecular weights and to measure water activity.

5.4.1   Boiling Point Elevation

If a small amount of nonvolatile solute is dissolved in a volatile solvent and the solution is very dilute to behave ideally, the lowered vapor pressure can be calculated from Eq. (5.25). As a result of lowered vapor pressure, the boiling point of solution is higher than that of the pure solvent.

As discussed before, chemical potentials of the volatile A for liquid and vapor phases are equal to each other at equilibrium as in Eq. (5.1):

$$\mu_A^L = \mu_A^V \tag{5.1}$$

The chemical potential of A in liquid phase is expressed as in Eq. (5.27):

$$\mu_A^L = \mu_A^{0L} + RT \ln P_A^0 + RT \ln X_A \tag{5.27}$$

At constant temperature the first two terms are constants and independent of composition. Therefore, they can be combined to simplify the equation:

$$\mu_A^L = \mu_A^{0L'} + RT \ln X_A \tag{5.29}$$

where $\mu_A^{0L'}$ is the chemical potential of pure liquid A.

At boiling point at 1 atm, $\mu_A^V = \mu_A^{0V'}$. Thus, Eq. (5.29) becomes:

$$\mu_A^{0V'} = \mu_A^{0L'} + RT \ln X_A \tag{5.30}$$

For a pure component A, the chemical potentials are identical with molar free energies. Thus:

$$\bar{G}_A^{0V} - \bar{G}_A^{0L} = RT \ln X_A \tag{5.31}$$

Remembering Eqs. (5.8) and (5.3) and writing them in terms of change in thermodynamic properties:

$$\left[\frac{\partial(\Delta G)}{\partial T} \right]_p = -\Delta S = \frac{\Delta G - \Delta H}{T} \tag{5.32}$$

This equation involves both the Gibbs free energy and temperature derivative of Gibbs free energy and it is more convenient to transform it so that only a temperature derivative appears. This can be achieved by first differentiating ($\Delta G/T$) with respect to temperature at constant pressure:

$$\left[\frac{\partial}{\partial T}\left(\frac{\Delta G}{T}\right)\right]_p = -\frac{\Delta G}{T^2} + \frac{1}{T}\left(\frac{\partial \Delta G}{\partial T}\right)_p \tag{5.33}$$

Using Eqs. (5.32) and (5.33), the Gibbs-Helmholtz equation is obtained:

$$\left[\frac{\partial}{\partial T}\left(\frac{\Delta G}{T}\right)\right]_p = -\frac{\Delta H}{T^2} \tag{5.34}$$

Substituting Eq. (5.31) into (5.34) and differentiating:

$$\bar{H}_A^{0V} - \bar{H}_A^{0L} = -RT^2\frac{d \ln X_A}{dT} \tag{5.35}$$

where ($\bar{H}_A^{0V} - \bar{H}_A^{0L}$) is the molar latent heat of vaporization (λ_v). Thus:

$$-d \ln X_A = \frac{\lambda_v}{RT^2}dT \tag{5.36}$$

Equation (5.36) can be integrated by using the limits of pure solvent ($X_A = 1$) at temperature T_0 to any arbitrary X_A and T values.

$$-\int_1^{X_A} d \ln X_A = \frac{\lambda_v}{R}\int_{T_0}^T \frac{dT}{T^2} \tag{5.37}$$

$$-\ln X_A = \frac{\lambda_v}{R}\left(\frac{1}{T_0} - \frac{1}{T}\right) = \frac{\lambda_v}{R}\left(\frac{T - T_0}{T T_0}\right) \tag{5.38}$$

Boiling point elevation ($T - T_0$) can be expressed as ΔT_B. When boiling point elevation is not too large, the product of T times T_0 can be replaced by T_0^2.

Considering the mole fraction of solute as X_B, the mole fraction of solvent is $X_A = 1 - X_B$

$\ln (1 - X_B)$ can be expanded in a power series as:

$$\frac{\lambda_v \Delta T_B}{RT_0^2} = X_B + \frac{1}{2}X_B^2 + \frac{1}{3}X_B^3 + \cdots \tag{5.39}$$

For dilute solutions where X_B is a very small fraction, higher order terms in Eq. (5.39) can be neglected. As a result:

$$\Delta T_B = \frac{RT_0^2}{\lambda_v}X_B \tag{5.40}$$

where

$$X_B = \frac{w_B M_A}{w_A M_B} \tag{5.41}$$

In which w_B and w_A are the masses of solute and solvent, respectively and M_B and M_A are the molecular weights of the solute and solvent, respectively. Substituting Eq. (5.41) into Eq. (5.40):

$$\Delta T_B = \frac{RT_0^2}{\lambda_v}\frac{w_B M_A}{w_A M_B} = \frac{RT_0^2}{l_v}\frac{w_B}{w_A M_B} \tag{5.42}$$

where l_v is the latent heat of vaporization per unit mass.

The term $\left({^{w_B}}/_{w_A M_B} \right)$ is expressed in terms of molality as $(m/1000)$. Then:

$$\Delta T_B = \frac{R T_0^2}{l_v} \frac{m}{1000} = K_B m \tag{5.43}$$

where K_B is the molal boiling point elevation constant.

This equation is valid for all nonelectrolytes in which equal numbers of moles are involved before and after the solution is formed. For electrolytes, Eq. (5.43) can be expressed as in Eq. (5.44):

$$\Delta T_B = i \frac{R T_0^2}{l_v} \frac{m}{1000} = i K_B m \tag{5.44}$$

where the i factor is the ratio of number of total moles after the solution to the number of moles before the solution. For nonelectrolytes $i = 1$.

To predict boiling point rise due to the solutes in the solution, an empirical rule known as Dühring's rule can be used. In this rule, a straight line is obtained if the boiling point of solution is plotted against the boiling point of pure water at the same pressure for a given concentration. For each concentration and pressure a different line is obtained.

A chart is given in Perry and Chilton's book (1973) to estimate the boiling point rise of a large number of common aqueous solutions used in chemical and biological processes.

Example 5.1.

(a) Determine the boiling temperature of 10% (w/w) NaCl solution under atmospheric pressure. Check the result also from the Dühring plot (Fig. E.5.1.1).

(b) By using the Dühring plot, estimate the boiling point of the same solution under a pressure of 47.39 kPa. What is the boiling point elevation at this pressure if at 47.39 kPa water boils at 80°C?

Data: Molecular weight of NaCl: 58.4 g/g-mole
Enthalpy of saturated vapor: 2676.1 kJ/kg at 100°C
Enthalpy of saturated liquid: 419.04 kJ/kg at 100°C
R, gas constant: 8.31434 kJ/kg-mole K.

Solution:

(a) For electrolytes such as NaCl, K_B can be calculated using Eq. (5.44):

$$\Delta T_B = \frac{i R T_0^2}{l_v} \frac{m}{1000} = i K_B m \tag{5.44}$$

$$K_B = \frac{R T_0^2}{l_v 1000} = \frac{8.3143 \, (373.15)^2}{(2676.1 - 419.04) \, 1000}$$

$$K_B = 0.513°C/m$$

Molality is the number of moles of solute in 1000 g of solvent. Molality of 10% sodium chloride solution can be calculated by using its molecular weight:

$$m = \left(\frac{10 \text{ g NaCl}}{90 \text{ g water}} \right) \left(\frac{1 \text{ gmole NaCl}}{58.4 \text{ g NaCl}} \right) (1000 \text{ g water}) = 1.9 \, m$$

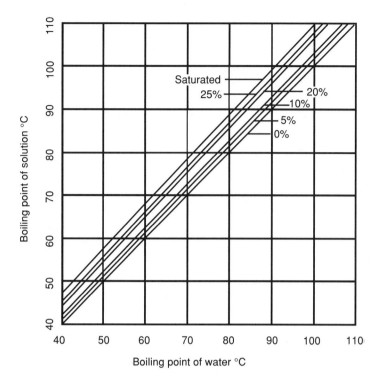

Figure E.5.1.1. Dühring chart for aqueous solutions of NaCl.

The i factor for NaCl is 2. Then,

$$\Delta T_B = i K_B m = (0.513)(1.9)(2) = 1.95°C$$

$$T_B = 100 + 1.95 = 101.95°C$$

From Fig. E.5.1.1, the boiling point of the solution is read as 103°C.

(b) At 47.39 kPa water boils at 80°C. From the Dühring plot, the boiling point of 10% NaCl solution for the pressure yielding boiling point of water at 80°C can be read as 83°C. Therefore, the boiling point elevation is:

$$\Delta T_B = 3°C$$

5.4.2   Freezing Point Depression

When equilibrium exists between solid A and its solution, the chemical potentials of A must be the same in both phases:

$$\mu_A^S = \mu_A^L \tag{5.45}$$

Using Eqs. (5.29) and (5.45):

$$\mu_A^S = \mu_A^L = \mu_A^{0L'} + RT \ln X_A \tag{5.46}$$

where μ_A^s and $\mu_A^{0L'}$ are the chemical potentials of pure solid and pure liquid A.
Since chemical potential is the molar free energy:

$$\frac{\bar{G}_A^{0S} - \bar{G}_A^{0L}}{T} = R \ln X_A \tag{5.47}$$

Differentiating Eq. (5.47) with respect to temperature at constant pressure yields:

$$\left[\frac{\partial}{\partial T}\left(\frac{\Delta \bar{G}}{T}\right)\right]_p = R\frac{\partial \ln X_A}{\partial T} \tag{5.48}$$

Substituting the Gibbs-Helmholtz equation [Eq. (5.34)] into Eq. (5.48):

$$-\left(\frac{\bar{H}_A^S - \bar{H}_A^{0L}}{RT^2}\right) = \frac{\partial \ln X_A}{\partial T} = \frac{\Delta H_{fus}}{RT^2} \tag{5.49}$$

where ΔH_{fus} is the latent heat of fusion. Integrating Eq. (5.49) by using the limits of mole fraction X_A at temperature T and $X_A = 1$ at the freezing point of component, T_0:

$$\int_{X_A}^{1} d\ln X_A = \int_{T}^{T_0} \frac{\Delta H_{fus}}{RT^2}dT \tag{5.50}$$

Assuming that ΔH_{fus} is independent of T:

$$-\ln X_A = \frac{\Delta H_{fus}(T_0 - T)}{RTT_0} \tag{5.51}$$

For small freezing point depression:

$$-\ln X_A = \frac{\Delta H_{fus}\Delta T_f}{RT_0^2} \tag{5.52}$$

where ΔT_f is the freezing point depression.

This equation has been derived for ideal solutions but it is applicable for nonideal solutions if the mole fraction of the solvent is very close to unity.

As discussed in Section 5.4.1, for dilute solutions $(-\ln X_A)$ can be represented by power series in X_B, the mole fraction of solute as shown in Eq. (5.39). For sufficiently low concentrations of solute, the second and higher order terms of this series are negligible. Then, Eq. (5.52) can be written as:

$$\Delta T_f = \frac{RT^{0^2}}{\Delta H_{fus}}X_B \tag{5.53}$$

where X_B is the same as defined in the boiling point elevation part:

$$X_B = \frac{w_B M_A}{w_A M_B} \tag{5.41}$$

in which w_B and w_A are the masses of solute and solvent, respectively and M_B and M_A are the molecular weights of the solute and solvent, respectively.

In freezing point depression, the term $\left(^{w_B}\!/_{w_A M_B}\right)$ is expressed in terms of molality as $(m/1000)$. Then Eq. (5.53) becomes:

$$\Delta T_f = \frac{RT^{0^2} M_A m}{1000 \Delta H_{fus}} = K_f m \tag{5.54}$$

where K_f is the molal freezing point depression constant.

Freezing point depression is useful for determining the molar mass of the solutes. These relations apply to ideal solutions. For electrolytes, Eq. (5.54) can be expressed as Eq. (5.55):

$$\Delta T_f = i \frac{RT^{0^2} M_A m}{1000 \Delta H_{fus}} = i K_f m \tag{5.55}$$

where i is the ratio of number of total moles after the solution to the number of moles before the solution. For nonelectrolytes $i = 1$.

Example 5.2. In a city, the administrative board of the municipality is discussing using salt (NaCl) or glycerol ($C_3H_8O_3$) for road treatment to eliminate ice. Assuming that you are one of the engineers taking part in the discussion, help them to decide which is the more effective type of antifreeze. Compare the performances of both types of solutions at the same concentration such as 10% (w/w).

Data: Latent heat of fusion for ice at $0°C$: 6028.5 kJ/kg-mole
Molecular weight of glycerol: 92 kg/kg-mole
Molecular weight of water: 18 kg/kg-mole
The gas constant, R: 8.3143 kJ/kg-mole·K

Solution:

Freezing point depression is calculated from Eq. (5.55):

$$\Delta T_f = i \frac{RT^{0^2} M_A m}{1000 \Delta H_{fus}} = i K_f m \tag{5.55}$$

$$K_f = \frac{RT^{0^2} M_A}{1000 \Delta H_{fus}} = \frac{(8.31434)(273.15)^2 (18)}{(1000)(6028.5)} = 1.85°C/m$$

For NaCl solution, molality of 10% sodium chloride solution is $1.9\,m$, from Example 5.1. The i factor for NaCl solution is 2. Then, freezing point depression is:

$$\Delta T_{f,\text{NaCl}} = (1.85)(1.9)(2) = 7.03°C$$

For a 10% glycerol solution molality is calculated as:

$$m = \left(\frac{10 \text{ kg glycerol}}{90 \text{ kg water}}\right)\left(\frac{1 \text{ kgmole glycerol}}{92 \text{ kg glycerol}}\right)(1000 \text{ kg water}) = 1.21m$$

The i factor for glycerol is 1.

$$\Delta T_{f,\text{gly}} = (1.85)(1.21)(1) = 2.24°C$$

Since $\Delta T_{f,\text{NaCl}}$ is greater than $\Delta T_{f,\text{gly}}$ for the same concentration, NaCl is a more effective antifreeze.

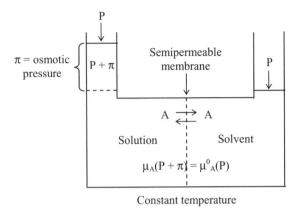

Figure 5.1   Osmometer showing equilibrium conditions. A represents solvent molecule.

5.4.3 Osmotic Pressure

Osmotic pressure is a colligative property that is closely related to the vapor pressure, freezing point, and boiling point. The activity of the solvent affects mostly osmotic pressure.

An osmotic pressure arises when two solutions of different concentrations are separated by a semipermeable membrane. Figure 5.1 shows the osmometer showing conditions at equilibrium. The solution and solvent are separated by a semipermeable membrane that allows only passage of solvent. Osmotic flow continues until the chemical potential of the diffusing component is the same on both sides of the barrier. The equilibrium occurs in an osmometer in which the extra pressure π necessary for equilibrium is produced by the hydrostatic pressure of a water column on the solution side.

At equilibrium there will be no net flow of solvent across the membrane so the chemical potential of pure solvent at pressure P must be equal to the chemical potential of solvent in solution at pressure $P + \pi$.

$$\mu_A(\text{solution}, P + \pi) = \mu_A(\text{solvent}, P) \tag{5.56}$$

At constant temperature and at 1 atm pressure:

$$\mu_A(\text{solution}, 1 + \pi) = \mu_A^0(\text{solvent}) \tag{5.57}$$

The chemical potential of A at 1 atm is decreased owing to the presence of solute in solution and it can be calculated as:

$$\Delta\mu_{\text{osmosis}} = \mu_A(\text{solution}, 1) - \mu_A^0(\text{solvent}) \tag{5.58}$$

Remembering Eq. (5.20):

$$d\mu_A = \bar{V}_A dP_A = RT\frac{dP_A}{P_A} \tag{5.20}$$

Integrating Eq. (5.20):

$$\Delta\mu_{\text{osmosis}} = \int_{1+\pi}^{1} \bar{V}_A dP = \int_{P_A}^{P_A^0} RT \frac{dP_A}{P_A} \tag{5.59}$$

$$\Rightarrow \bar{V}_A \pi = RT \ln\left(\frac{P_A^0}{P_A}\right) \tag{5.60}$$

Generally, molar volume of solvent in solution, $\bar{V}_A$, can be approximated by the molar volume of pure liquid, $\bar{V}_A^0$. For ideal solutions, Eq. (5.60) becomes:

$$\bar{V}_A^0 \pi = -RT \ln X_A \tag{5.61}$$

For dilute solutions $\ln X_A$ can be approximated as:

$$\ln X_A = \ln(1 - X_B) \cong -X_B = -\frac{n_B}{n_A + n_B} \cong -\frac{n_B}{n_A} \tag{5.62}$$

Substituting Eq. (5.62) into Eq. (5.61):

$$\pi = \frac{RT}{n_A \bar{V}_A^0} n_B \tag{5.63}$$

where π is osmotic pressure (Pa), V is the volume of solution (m^3), and n_B is the number of moles of solute in solution.

In dilute solutions $n_A \bar{V}_A^0$ is equal to the volume of solution since the volume of the solute is negligible. Therefore:

$$\pi V = n_B RT \tag{5.64}$$

$$\Rightarrow \pi = cRT \tag{5.65}$$

where R is the gas constant (8314.34 m$^3\cdot$ Pa/kg-mole $\cdot$ K) and c is the concentration of solute (kg-mole/m^3).

5.5   EQUILIBRIA IN NONIDEAL SYSTEMS—FUGACITY AND ACTIVITY

As discussed earlier, for ideal systems, the change in chemical potential is expressed as in Eq. (5.21):

$$\int_{\mu_A^0}^{\mu_A} d\mu_A = \int_{1}^{P_A} RT \frac{dP_A}{P_A} \tag{5.21}$$

Chemical potential of component A at pressure P_A is expressed as in Eq. (5.22):

$$\mu_A = \mu_A^0 + RT \ln P_A \tag{5.22}$$

These equations are not valid for real gases unless $P \to 0$. Lewis proposed a new function called fugacity (f) to make these equations applicable to real systems. Fugacity is defined as the escaping tendency of a component in solution. It can be considered as corrected partial pressure.

A dimensionless parameter called fugacity coefficient, γ_f, is defined to calculate the relation between the fugacity and pressure:

$$f_A = \gamma_{fA} P_A \tag{5.66}$$

For $\gamma_{f_A} = 1$, the following equation, which is analogous to Eq. (5.21), can be written:

$$\int_{\mu_A^0}^{\mu_A} d\mu_A = \int_{f_A^0}^{f_A} RT \frac{df}{f} \tag{5.67}$$

Integrating between the given state and standard state, the following equation can be obtained:

$$\mu_A = \mu_A^0 + RT \ln \frac{f_A}{f_A^0} \tag{5.68}$$

The ratio of fugacity (f_A) to the fugacity in standard state (f_A^0) is defined as the activity (a_A):

$$a_A = \frac{f_A}{f_A^0} \tag{5.69}$$

Then:

$$\mu_A = \mu_A^0 + RT \ln a_A \tag{5.70}$$

As dilution increases, the solvent in solution approaches the ideal behavior specified by Raoult's law. For this case, the standard state of component A in solution is taken to be pure liquid at 1 atm and at the given temperature activity can be expressed as:

$$a_A = \frac{f_A}{f_A^0} \approx \frac{P_A}{P_A^0} \tag{5.71}$$

where P_A^0 is the vapor pressure of pure A at 1 atm. Thus, Raoult's law becomes:

$$a_A = \frac{P_A}{P_A^0} = X_A \tag{5.72}$$

For an ideal solution, or for any solution in the limit as X_A approaches 1, X_A is equal to a_A.

5.6 WATER ACTIVITY

Consider a food system enclosed in a container. All the components in the food are in thermodynamic equilibrium with each other in both the adsorbed and vapor phases at constant temperature. Considering moisture within the food system and the vapor in the headspace, their chemical potentials are equal to each other:

$$\mu_w^{food} = \mu_w^{vapor} \tag{5.73}$$

The activity of water in foods can be expressed as in Eq. (5.71):

$$a_w = \frac{f_w}{f_w^0}$$

If the water activity is expressed in terms of pressures, the following equation is obtained:

$$a_w = \gamma_{fw} \frac{P_w}{P_w^0} \tag{5.74}$$

Fugacity coefficients of water vapor in equilibrium with saturated liquid are above 0.90 at temperatures between 0.01°C and 210°C and pressure values between 611 Pa and 19.1 × 10⁵ Pa (Rizvi, 2005). Therefore, it can be approximated to 1, which means negligible deviation from ideality. Thus, water activity in foods can be expressed as:

$$a_w = \frac{P_w}{P_w^0} \tag{5.75}$$

Therefore, water activity can be defined as the ratio of the vapor pressure of water in the system to the vapor pressure of pure water at the same temperature. It can also be expressed as the equilibrium relative humidity (ERH) of the air surrounding the food at the same temperature.

Water activity is an important property in food systems. Most chemical reactions and microbiological activity are controlled directly by the water activity. In food science, it is very useful as a measure of the potential reactivity of water molecules with solutes.

Water activity deviates from Raoult's law at higher solute concentrations. There are many reasons for this deviation. The variation of solute and solvent molecule size results in a change of intermolecular forces between the molecules, leading them to behave as nonideal. The intermolecular forces between solvent molecules, solute–solute and solute–solvent interactions and solvation effects cause the solution to behave nonideally. If the solute completely dissociates into ions when dispersed in a solvent it causes nonideality. The presence of insoluble solids or porous medium cause a capillary action that decreases vapor pressure and shows deviation from ideal behavior. The reasons for nonideal behavior (deviation from Raoult's law) are explained in detail in Rahman (1995).

5.7   PREDICTION OF WATER ACTIVITY

Raoult's law is the basic equation for determination of water activity of ideal solutions. According to Raoult's law, water activity is equal to the mole fraction of water in the solution:

$$a_w = \frac{X_w^w}{X_w^w + \left(M_w/M_s\right) X_s^w} \tag{5.76}$$

where X^w is the mass fraction and M is the molecular weight. The subscript w is for water and s for solute.

As mentioned in Section 5.6, Raoult's law is not valid for macromolecular solute due to the very low molecular weight ratio of water and solute. If the solute ionizes in solution Raoult's law can be written as:

$$a_w = \frac{X_w^w}{X_w^w + \psi \left(M_w/M_s\right) X_s^w} \tag{5.77}$$

where ψ is the degree of ionization.

The Gibbs-Duhem equation was used to produce series of predictive models of water activity. The Gibbs-Duhem equation can be used to describe the activity of solutions composed of N components (Cazier & Gekas, 2001).

$$\sum_{i=1}^{N} n_i d(\ln a_i) = 0 \tag{5.78}$$

If we consider water solution and the activity is given as a function of molality, m_i and activity coefficient, γ_i:

$$a_i = m_i \gamma_i \tag{5.79}$$

Substituting Eq. (5.79) into Eq. (5.78) and integrating gives:

$$\ln a_w = -\frac{1}{55.5} \sum_{i=1}^{N} \int_{m_1} m_i d \left[\ln (m_i \gamma_i) \right] \tag{5.80}$$

Thus, a distribute integration is possible because the molality is not a function of activity:

$$\ln a_w = -\frac{1}{55.5} \sum_{i=1}^{N} m_i d \left(1 + \ln \gamma_i \right) \tag{5.81}$$

In this system, the activity coefficients, γ_i, for any solution cannot be known from standard tables. Moreover, it is not easy to solve such an equation analytically. Therefore, Ross (1975) simplified the model to obtain a simple and usable model. According to the Ross model, the interactions between diluted components are neglected. In the binary case, the following equation was obtained:

$$a_w = 1 - \frac{m_s}{55.5} (1 + \ln \gamma_s^0) \tag{5.82}$$

In the case of multicomponent dilute solutions, water activity can be approximated as equivalent to the product of the standard binary activity, a_i^0 of each component.

$$a_w \approx \prod_{i=1}^{N} a_i^0 \tag{5.83}$$

The simplifications used in Ross model restricts its use to very dilute solutions in which various interactions cancel each other or are of no importance. Caurie (1985) proposed a modification to make a simpler way to evaluate the Gibbs-Duhem equation based on the Ross model without such losses. However, the proposed solution is not very suitable in a high concentration range.

$$a_w \approx \prod_{i=1}^{N} a_i^0 - \left(\frac{N \prod_{i \neq j}^{N} m_i m_j}{55.5^2} + \frac{(N+1) \prod_{i=1}^{N} m_i}{55.5^3} \right) \tag{5.84}$$

Example 5.3. The ancient Egyptians preserved meat in the form of mummification. The embalmers used natron, which is a mixture of sodium carbonate with 10 molecules of water ($Na_2CO_3 \cdot 10H_2O$) and 10% to 30% sodium chloride (NaCl). During the embalming process, it may be assumed that after a long time, the body reaches equilibrium with the saturated solution of natron. The water activity of natron was recently measured as 0.71. This means that if a fresh body with water activity of about 0.99 is put in this solution, the level at which equilibrium is reached is low enough to inhibit almost all kinds of microbial growth.

Similarly, in a curing process, meat is plunged in NaCl solution with a_w value of 0.75 and allowed to reach equilibrium with the solution. If a *Streptococcus faecalis* cell with an internal concentration of 0.8 osmole/L cell solution contaminate the meat, how much osmotic pressure does the cell experience at room temperature?

Data: R gas constant: 8314.34 m³Pa/kg-mole.K
Molecular weight of NaCl: 58.4 kg/kg-mole
Molecular weight of water: 18 kg/kg-mole
Density of the NaCl solution: 1197 kg/m³ at room temperature.

Hints:

(a) Osmole is the unit for the amount of substance that dissociates in solution to form 1 mole of osmotically active particles, e.g., 1 mole of glucose, which is not ionizable, forms 1 osmole of solute, but 1 mole of sodium chloride forms 2 osmoles of solute.
(b) Ignore the deviation of water activity from Raoult's law at high solute concentrations.
(c) Assume that the cell membrane act as a semipermeable membrane allowing only the passage of water.

Solution:

Assuming Raoult's law is valid, Eq. (5.72) can be written for water as:

$$a_w = X_w = \frac{n_{\text{solution}} - n_{\text{solute}}}{n_{\text{solution}}} = 0.75$$

$$\Rightarrow 0.25 n_{\text{solution}} = n_{\text{solute}}$$

In 1 mole of solution, there is 0.75 mole of water and 0.25 mole of solute. One mole of NaCl yields 2 osmoles of solute. Therefore, 0.25 mole of solute is obtained from 0.125 mole NaCl. By using the number of moles and molecular weight of water and NaCl, the mass of water and NaCl is calculated as 13.5 g and 7.3 g, respectively. Thus, 1 mole of solution is 20.8 g.

The volume of NaCl solution is:

$$V = 20.8 \text{ g} \left(\frac{1 \text{ kg}}{1000 \text{ g}} \right) \left(\frac{1 \text{ m}^3}{1197 \text{ kg}} \right) \left(\frac{10^6 \text{ ml}}{1 \text{ m}^3} \right) = 17.4 \text{ ml solution}$$

The concentration of NaCl solution (c_1) is found as:

$$c_1 = \frac{0.25 \text{ kg-mole solute}}{17.4 \text{ ml solution}} = 0.0144 \frac{\text{kgmole solute}}{\text{ml solution}} = 14.4 \frac{\text{kgmole solute}}{\text{L solution}}$$

The concentration of cell solution (c_2) is given as 0.8 kg-mole/L
Then:

$$c = c_1 - c_2 = 14.4 - 0.8 = 13.6 \text{ kg-mole/L} = 0.0136 \text{ kg-mole/cm}^3$$

$$\pi = cRT \tag{5.65}$$

$$\pi = \left(13.6 \frac{\text{mol}}{\text{L}} \right) \left(\frac{1000 \text{ L}}{1 \text{ m}^3} \right) \left(8314.34 \frac{\text{m}^3 \cdot \text{Pa}}{\text{kgmole·K}} \right) (298.15 \text{ K}) = 33.71 \times 10^6 \text{ kPa}$$

Example 5.4. NaCl, sucrose or the NaCl-sucrose solutions are commonly used for osmotic dehydration of potatoes.

(a) Estimate the water activity of 20% sucrose solution.
(b) Estimate the water activity of 20% NaCl solution
(c) Estimate the water activity of solution containing 10% NaCl and 10% sucrose.
(d) Which solution do you think will be more efficient for osmotic dehydration of potatoes having water activity of 0.942?

Data: Molecular weight of water: 18 kg/kg-mole
Molecular weight of NaCl: 58.44 kg/kg-mole
Molecular weight of sucrose: 342 kg/kg-mole
NaCl ionizes and its maximum degree of ionization, ψ: 2.

Solution:

(a)

$$X_s^w = 0.20$$

$$X_w^w = 1 - X_s^w = 0.80$$

Using Eq. (5.76):

$$a_w = \frac{X_w^w}{X_w^w + \left(M_w / M_s\right) X_s^w} \qquad (5.76)$$

$$a_w = \frac{0.80}{0.80 + \left(18 / 342\right) 0.20}$$

$$a_w = 0.987$$

(b) Since NaCl ionizes in solution, Eq. (5.77) is used:

$$a_w = \frac{X_w^w}{X_w^w + \psi \left(M_w / M_s\right) X_s^w} \qquad (5.77)$$

$$a_w = \frac{0.80}{0.80 + 2 \left(18 / 58.44\right) 0.20}$$

$$a_w = 0.867$$

(c) For the solution containing 10% NaCl and 10% sucrose:

$$a_w = \frac{0.80}{0.80 + 1 \left(18 / 342\right) 0.10 + 2 \left(18 / 58.44\right) 0.10}$$

$$a_w = 0.923$$

(d) When 20% sucrose solution is used, potatoes will adsorb moisture. Using a 20% NaCl solution is preferred to the solution containing 10% NaCl and 10% sucrose.

5.8   WATER ACTIVITY MEASUREMENT METHODS

A variety of methods for measurement of water activity have been used and reported in the literature (Bell and Labuza, 2000; Nunes, Urbicain & Rotstein, 1985; Troller, 1983). In general, for measuring water activity of foods methods based on colligative properties, isopiestic transfer, hygroscopicity of salts, and hygrometers can be used.

5.8.1   Measurements Based on Colligative Properties

Water activity of foods can be determined either by measuring the vapor pressure of water in food directly or by using freezing point depression. Colligative properties such as osmotic pressure and boiling point elevation have not been used so far for food systems (Rizvi, 2005).

5.8.1.1   Water Activity Determination by Vapor Pressure Measurement

This method gives a direct measure of vapor pressure exerted by sample. Water activity is calculated from the ratio of vapor pressure of sample to that of pure water at the same temperature.

Figure 5.2 shows a schematic diagram for this method. According to this method, a sample weighing 10 to 50 g is put into a sample flask and sealed on to the apparatus. The air space in the apparatus is evacuated. After the vacuum source is isolated and equilibration for 30 to 50 min, the pressure exerted by the sample is recorded as Δh_1. The level of oil in the manometer will change by the vapor pressure exerted by the sample. The sample flask is excluded from the system and the desiccant flask is opened. Water vapor is removed by sorption onto $CaSO_4$ and the pressure exerted by volatiles and gases are indicated by Δh_2 after equilibration. Then, water activity can be calculated using Eq. (5.85):

$$a_w = \frac{(\Delta h_1 - \Delta h_2)}{P^0} \rho g \tag{5.85}$$

Temperature must be constant during measurement. If the temperatures of the sample (T_s) and vapor space in the manometer (T_m) are different, water activity is corrected as:

$$a_w = \frac{(\Delta h_1 - \Delta h_2)}{P^0} \left(\frac{T_s}{T_m}\right) \rho g \tag{5.86}$$

Troller (1983) modified the set up by replacing the oil manometer with a capacitance manometer. As a result, a more compact devise in which the temperature control is less problematic, is obtained.

The ratio of volume of sample to that of vapor space should be large enough to minimize the changes in water activity due to loss of water by vaporization. Oil, having a low density and low vapor pressure, should be used as the manometer fluid. Volatiles other than water may contribute to the pressure exerted by the food. If there is microbial growth in the system, gas evolution prevents vapor pressure equilibrium.

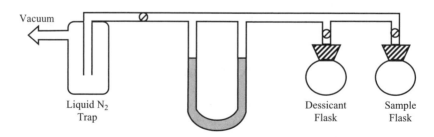

Vacuum

Liquid N₂ Trap   Dessicant Flask   Sample Flask

Figure 5.2   Schematic diagram for vapor pressure manometric method.

5.8.1.2   Water Activity Determination by Freezing Point Depression

The determination of water activity by freezing point depression is very accurate at water activities above 0.85 (Rizvi, 2005). However, this method is applicable only to liquid foods and gives water activity values at freezing point rather than at room temperature. This method is suitable for materials having large quantities of volatile substances which may cause error in vapor pressure measurement and in electric hygrometers.

In two-phase systems (ice and solution) at equilibrium the vapor pressure of water as ice crystals and the interstitial concentrated solution are the same and water activity depends only on temperature. Thus, water activity of solution at a certain temperature below freezing can be expressed as:

$$a_w = \frac{\text{Vapor pressure of ice}}{\text{Vapor pressure of liquid water}}$$

(5.87)

5.8.2   Measurements Based on Isopiestic Transfer

Water activity measurement in this method is achieved by equilibration of water activities of two materials in a closed system. Mostly, microcrystalline cellulose is used as the reference adsorption substrate. Sample and predried microcrystalline cellulose are placed in the vacuum type desiccator. The desiccator is closed and evacuated for about 1 min and then held at constant temperature for 24 hours. After 24 hours, the vacuum on the desiccator is slowly released. The cellulose is reweighed and the change in weight is recorded. The moisture content is calculated and water activity is determined from a standard cellulose isotherm that was previously prepared with sulfuric acid–water mixtures as the medium. This method is not recommended for samples that are susceptible to foaming such as protein solutions during evacuation of desiccators. Such types of samples may be degassed prior to evacuation. Microbial growth must be prevented by the addition of K-sorbate and by using aseptic techniques.

The reasons for selection of microcrystalline cellulose as a reference can be summarized as follows (Spiess & Wolf, 1987):

1. It is stable in the temperature change of -18 to $80°C$ with little changes in sorption characteristics.
2. It is stable in its sorption properties after two to three repeated adsorption and desorption cycles.
3. Its sorption isotherm is in sigmoid shape and its sorption model is known.
4. It is available as a standard biochemical analytical agent.

5.8.3   Measurements Using Hygrometers

In this method, the sample is equilibrated with air in a closed vessel and then the relative humidity of the air is determined by using a hygrometer. Many hygrometric instruments work on the principle of measuring wet and dry bulb temperature, dew point, change in length of material, and electrical resistance or capacitance of salt (Rahman, 1995).

The relative humidity can be determined by measuring the wet and dry bulb temperatures of air in equilibrium with food. This method has been used primarily to determine the relative humidity of large storage atmospheres and commercial dehydrators (Troller & Christian, 1978).

In dew point temperature measuring instruments, an air stream in equilibrium with the sample is allowed to condense at the surface of a cold mirror. The relative humidity can be determined from the measured dew point temperature.

In mechanical hygrometers, relative humidity of the environment is measured by using dimensional changes in a material. The most common one is the hair type hygrometer. These types of hygrometers are not very sensitive.

Electrical resistance or capacitance hygrometers are based on measurement of the conductivity of salt solution that is in equilibrium with the air. Usually LiCl is used for this purpose. These types of hygrometers provide rapid and reliable means of measuring water activity.

5.8.4   Measurements Based on Hygroscopicity of Salts

Each salt has its own characteristic or critical relative humidity transition point. It will remain dry if the relative humidity of the surrounding air is lower than that of the critical. It will show a wet short line that can be seen under a lens if the air relative humidity is higher than the critical. Using this property of salt, it is possible to observe the change in the color of salt enclosed over the sample such as silica gel, which changes color from blue at low humidity through lilac to pink at high humidity. Thus, it gives an approximate estimate of ERH from the color matching against a calibrated series of standard colors.

5.9   EFFECTS OF TEMPERATURE ON WATER ACTIVITY

The temperature dependence of water activity can be described by the Clasius-Clapeyron equation if the isosteric heat and water activity values at one temperature are known (Rizvi, 2005).

$$\ln \frac{a_{w2}}{a_{w1}} = \frac{q_{st}}{R} \left(\frac{1}{T_1} - \frac{1}{T_2} \right) \tag{5.88}$$

where q_{st} is net isosteric heat of sorption or excess heat of sorption, and a_{w1} and a_{w2} are the water activities at temperatures T_1 and T_2.

In Eq. (5.88), it is assumed that moisture content of the system remains constant and the isosteric heat of sorption does not change with temperature.

5.10   EFFECTS OF PRESSURE ON WATER ACTIVITY

The effects of pressure on adsorption isotherm are relatively small and negligible at reasonable pressure levels. The variation of water activity with pressure at constant water activity is given by Okos, Narsimhan, Singh, and Weitnauer, (1992) as:

$$\ln \frac{a_{w2}}{a_{w1}} = \frac{M W_w}{\rho R T} (P_2 - P_1) \tag{5.89}$$

where a_{w1} and a_{w2} are water activities at total pressures P_1 and P_2 (Pa), R is the gas constant (8314 m^3 Pa/kg-mole K), and T is temperature (K).

5.11 ADJUSTMENT OF WATER ACTIVITY AND PREPARATION OF MOISTURE SORPTION ISOTHERMS

The simplest method to obtain the sorption data of foods is storing a weighed sample in an enclosed container maintained at a certain relative humidity, at constant temperature, and reweighing it after equilibrium is reached. Theoretically, at equilibrium water activity of the sample is the same as that of the surrounding environment. However, in practice a true equilibrium is never attained because that would require an infinitely long period of time. Therefore, the sample is weighed from time to time during equilibration. When the difference between successive weights of the sample becomes less than the sensitivity of the balance being used, it is accepted that equilibrium is reached. The moisture content of the sample is then determined.

The desired relative humidity environments can be generated by using saturated salt solutions, sulfuric acid, or glycerol. Table 5.1 shows the water activity of selected saturated salt solutions at different temperatures. Although saturated salt solutions are commonly used, they provide only discrete water activity values at any given temperature. Water activity of most salt solutions decreases with an increase in temperature owing to their increased solubility and their negative heats of solution.

Water activity of aqueous sulfuric acid solutions at different concentrations and temperatures are given in Table 5.2. The main disadvantage of using sulfuric acid solutions is corrosion.

Water activities of glycerol solutions at different concentrations at 20°C are given in Table 5.3. When glycerol solutions are used, the range of water activities is narrower as compared to those in the case of saturated salt and sulfuric acid solutions (Tables 5.1–5.3).

Table 5.1 Water Activities of Selected Saturated Salt Solutions at Different Temperatures[a]

Salts	Water Activity						
	5°C	10°C	20°C	25°C	30°C	40°C	50°C
Potassium hydroxide	0.143	0.123	0.093	0.082	0.074	0.063	0.057
Sodium hydroxide	—	—	0.089	0.082	0.076	0.063	0.049
Lithium chloride	0.113	0.113	0.113	0.113	0.113	0.112	0.111
Potassium acetate	—	0.234	0.231	0.225	0.216	—	—
Magnesium chloride	0.336	0.335	0.331	0.328	0.324	0.316	0.305
Sodium iodide	0.424	0.418	0.397	0.382	0.362	0.329	0.292
Potassium carbonate	0.431	0.431	0.432	0.432	0.432	—	—
Magnesium nitrate	0.589	0.574	0.544	0.529	0.514	0.484	0.454
Potassium iodide	0.733	0.721	0.699	0.689	0.679	0.661	0.645
Sodium nitrate	0.786	0.757	0.725	0.709	0.691	0.661	0.645
Sodium chloride	0.757	0.757	0.755	0.753	0.751	0.747	0.744
Ammonium chloride	—	0.806	0.792	0.786	0.779	—	—
Potassium bromide	0.851	0.838	0.817	0.809	0.803	0.794	0.790
Ammonium sulfate	0.824	0.821	0.813	0.810	0.806	0.799	0.792
Potassium chloride	0.877	0.868	0.851	0.843	0.836	0.823	0.812
Potassium nitrate	0.963	0.960	0.946	0.936	0.923	0.891	0.848
Potassium sulfate	0.985	0.982	0.976	0.973	0.970	0.964	0.958

[a]From Greenspan (1977).

Table 5.2 Water Activities of Sulfuric Acid Solutions at Different Concentrations and Temperatures[a]

H_2SO_4 (%)	Density at 25°C (g/cm³)	Water Activity						
		5°C	10°C	20°C	25°C	30°C	40°C	50°C
5	1.0300	0.9803	0.9804	0.9806	0.9807	0.9808	0.9811	0.9814
10	1.0640	0.9554	0.9555	0.9558	0.9560	0.9562	0.9565	0.9570
15	1.0994	0.9227	0.9230	0.9237	0.9241	0.9245	0.9253	0.9261
20	1.1365	0.8771	0.8779	0.8796	0.8805	0.8814	0.8831	0.8848
25	1.1750	0.8165	0.8183	0.8218	0.8235	0.8252	0.8285	0.8317
30	1.2150	0.7396	0.7429	0.7491	0.7521	0.7549	0.7604	0.7655
35	1.2563	0.6464	0.6514	0.6607	0.6651	0.6693	0.6773	0.6846
40	1.2991	0.5417	0.5480	0.5599	0.5656	0.5711	0.5816	0.5914
45	1.3437	0.4319	0.4389	0.4524	0.4589	0.4653	0.4775	0.4891
50	1.3911	0.3238	0.3307	0.3442	0.3509	0.3574	0.3702	0.3827
55	1.4412	0.2255	0.2317	0.2440	0.2502	0.2563	0.2685	0.2807
60	1.4940	0.1420	0.1471	0.1573	0.1625	0.1677	0.1781	0.1887
65	1.5490	0.0785	0.0821	0.0895	0.0933	0.0972	0.1052	0.1135
70	1.6059	0.0355	0.0377	0.0422	0.0445	0.0470	0.0521	0.0575
75	1.6644	0.0131	0.0142	0.0165	0.0177	0.0190	0.0218	0.0249
80	1.7221	0.0035	0.0039	0.0048	0.0053	0.0059	0.0071	0.0085

[a]From Ruegg, M. Calculation of the activity of water in sulfuric acid solutions at various temperatures. *Lebensmittel Wissenschaft und Technologie, 13*, 22–24. Copyright © (1980) with permission from Elsevier.

The main disadvantages of preparation of moisture sorption isotherms with this method are the long equilibration time and the risk of mold or bacterial growth at high relative humidity. To decrease equilibration time, air inside the enclosure can be circulated and surface area of the sample can be increased. At high relative humidity values, storage should be at low temperature to prevent microbial

Table 5.3 Water Activities of Glycerol Solutions at 20°C[a]

Concentration (kg/L)	Refractive Index	Water Activity
0.2315	1.3602	0.95
0.3789	1.3773	0.90
0.4973	1.3905	0.85
0.5923	1.4015	0.80
0.6751	1.4109	0.75
0.7474	1.4191	0.70
0.8139	1.4264	0.65
0.8739	1.4329	0.60
0.9285	1.4387	0.55
0.976	1.4440	0.50

[a]From Grover and Nicol (1940).

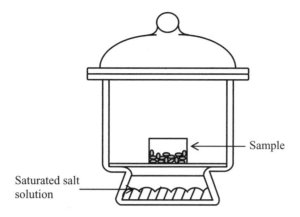

Figure 5.3 Desiccator method.

growth. To prevent mold or bacterial growth, use of aseptic techniques are suggested. In addition, toluene or potassium sorbate can be used (Rahman, 1995).

Desiccators can be used for preparation of sorption isotherms. In the desiccator method, saturated salt solutions, sulfuric acid or glycerol solutions are put into the bottom of desiccators (Fig. 5.3).

Although the desiccator method is very commonly used for water activity determination and preparation of sorption isotherms, there are some errors arising from this method that were recently discussed by Lewicki and Pomaranska-Lazuka (2003). According to their study, it was shown that error comes from the disturbance of equilibrium caused by opening the desiccators, taking the sample, and closing it again. These disturbances cause adsorption of water from the surrounding air by samples with low water activities and desorption of water from samples having high water activities. If desorption occurs, the results are not affected significantly since desorption occurs slowly. However, if adsorption occurs, water activity is affected significantly since adsorption is a fast process. To minimize these errors the following recommendations were given (Lewicki & Pomaranska-Lazuka, 2003):

1. When multiple samples are used in one desiccator, weighting bottles with covers should be used. When the desiccator is opened, all weighting bottles should be closed and then samples should be weighed.
2. If weighting bottles cannot be used, water activity of the sample should be measured each time after it is weighed.
3. The type of sample holders used and how water activities of samples are measured should always be specified.

A moisture sorption isotherm describes the relationship between water activity and the equilibrium moisture content of a food product at constant temperature. It is also called the equilibrium moisture content curve.

Equilibrium moisture content ($X*$) is the moisture content of a substance at equilibrium with a given partial pressure of the vapor. It is used to describe the final moisture content that will be reached during drying. Free moisture content is the moisture content in a substance in excess of equilibrium moisture content ($X - X*$). Free moisture can be removed by drying under the given percent relative humidity. The moisture content data can be given in a dry or wet basis. Moisture content is in a dry basis if it is expressed as the ratio of the amount of moisture in the food to the amount of dry solid (kg of moisture/kg of dry solid). If the moisture content of a sample is described as the ratio of the amount of water in the food to the total amount of wet solid (kg of moisture/kg of wet solid), it is in a wet basis. Moisture content is usually given in a wet basis to describe the composition of the food material. It is more common to use moisture content in a dry basis to describe moisture changes during drying.

The sorption isotherm is useful to determine the shelf life and to assess the background of operations such as drying, conditioning, mixing, packaging, and storage. The isotherm also gives information about the specific interaction between water and the product since it directly relates the thermodynamic potential (Gibbs free energy) of water in the system to its mass fraction (Van Den Berg, 1984).

It is necessary to adjust the water activity of food samples to a range of values in order to obtain the sorption data. The two principal techniques used for the adjustment of water activity are the integral and differential methods. In the integral method, several samples are prepared and each is placed under a controlled relative humidity environment simultaneously. The moisture contents of the samples are measured after constant weight is attained. In differential method, a single sample is placed under successively increasing or decreasing relative humidity environments. Moisture content is measured after each equilibration. The differential method has the advantage of using only a single sample. As a result, the error coming from the sample variation is eliminated. However, since equilibration can take several days, the sample may undergo various degenerative changes. The integral method avoids this problem because each sample is discarded after appropriate measurement is made.

Sorption isotherms of food materials are generally in sigmoid shape (type II) (Fig. 5.4). The effects of Raoult's law, capillary effects, and surface–water interactions are important in sorption curves and they are additive.

Type I isotherm is observed in pure crystalline sugar. It shows very little moisture gain up to a water activity of 0.7 to 0.8 since the only effect of water is hydrogen bonding to the –OH groups present on the surface of the crystal (Labuza, 1984). That is, surface effect is important, which means grinding the sugar to smaller particles will increase the moisture content at low water activity values. As the water activity is increased, water begins to penetrate into the crystal, causing dissociation of sugar–sugar interactions and a solution is obtained. At this stage, the effect of Raoult's law is important.

The type III isotherm is observed in the case of anticaking agents. In these types of materials, binding energy is so large that water activity is depressed while water is absorbed. When all the binding sites are filled, the increase in moisture content causes water activity to increase drastically.

5.11.1 Hysteresis

Sorption isotherms can be generated from an adsorption process (starting from a dry system having a zero water activity) or a desorption process (starting with a wet system having a water activity value of 1). The difference between these curves is defined as hysteresis (Fig. 5.5). Hysteresis is observed in most hygroscopic foods. In Fig. 5.5, in region A, water is tightly bound. In region B, the water is less tightly held and usually present in small capillaries, and in region C, water is free or loosely held in large capillaries (Fortes and Okos, 1980).

Desorption isotherms usually give higher moisture content than adsorption isotherms. The composition of the product, its temperature, storage time, drying temperature, and the number of successive adsorption and desorption affect hysteresis.

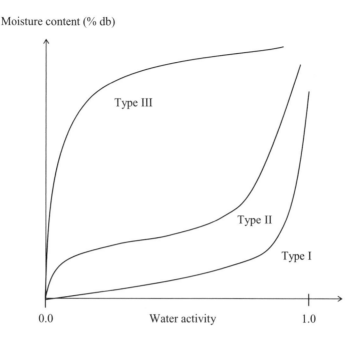

Figure 5.4 Classification of standard moisture sorption isotherms.

The proposed interpretations for sorption hysteresis can be classified as hysteresis on porous solids (theories based on capillary condensation), hysteresis on nonporous solids (based on surface impurities, partial chemisorption, or phase change), and hysteresis on nonrigid solids (based on changes in structure that result in hindered penetration and exit of the adsorbate) (Kapsalis, 1987).

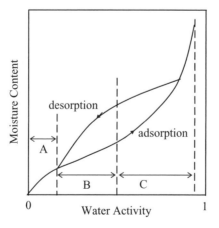

Figure 5.5 Sorption isotherm for typical food product showing the hysteresis.

5.11.2 Isotherm Models

There are some semiempirical equations with two or three fitting parameters to describe moisture sorption isotherms. The most common ones used in describing sorption on foods are the Langmuir equation, the Brunauer-Emmett-Teller (BET) equation, Oswin model, Smith model, Halsey model, Henderson model, Iglesias-Chirife equation, Guggenheim-Anderson-de Boer (GAB) model, and Peleg model.

(a) Langmuir Equation. Langmuir (1918) proposed the following physical adsorption model on the basis of unimolecular layers with identical, independent sorption sites:

$$a_w \left(\frac{1}{M_w} - \frac{1}{M_0} \right) = \frac{1}{CM_0} \tag{5.90}$$

where C is a constant and M_0 is the monolayer sorbate content.

(b) Brunauer-Emmett-Teller (BET) Equation. The BET equation, which is the most widely used model in food systems, was first proposed by Brunauer, Emmett, and Teller (1938). It holds at water activities from 0.05 to 0.45. It is an effective method for estimating the amount of bound water to specific polar sites in dehydrated food systems.

$$M_w = \frac{M_0 \, C \, a_w}{(1 - a_w)[1 + (C - 1) \, a_w]} \tag{5.91}$$

where M_0 is the monolayer moisture content and C is the energy constant related to the net heat of sorption. Monolayer moisture content represents the moisture content at which water attached to each polar and ionic groups starts to behave as a liquid-like phase.

(c) Oswin Model. An empirical model that is a series expansion for sigmoid shaped curves was developed by Oswin (1946):

$$M_w = C \left[\frac{a_w}{1 - a_w} \right]^n \tag{5.92}$$

where C and n are constants.
The Oswin equation was used to relate moisture content of nonfat dry milk and freeze-dried tea up to water activity of 0.5 (Labuza, Mizrahi, & Karel, 1972).

(d) Smith Model. Smith (1947) developed an empirical model to describe the final curved portion of water sorption isotherm of high molecular weight biopolymers. This model can be written as:

$$M_w = C_1 + C_2 \ln (1 - a_w) \tag{5.93}$$

where C_1 and C_2 are constants.
This equation could be used between water activity of 0.5 to 0.95 in the case of wheat desorption (Rahman, 1995).

(e) Halsey Model. The sorption behavior of starch-containing foods has been well described by the Halsey equation (Crapiste and Rotstein, 1982). According to the Halsey (1948) model, water activity is described as:

$$M_w = M_0 \left(-\frac{A}{RT \ln a_w} \right)^{1/n} \tag{5.94}$$

where A and n are constants and M_0 is monolayer moisture content.

Since the use of the RT term does not eliminate the temperature dependence of A and n, the Halsey equation was modified to the following form by Iglesias and Chirife (1976):

$$M_w = \left(-\frac{C}{\ln a_w} \right)^{1/n} \tag{5.95}$$

where C and n are constants.

(f) Henderson Model. The Henderson (1952) model is a commonly used model and can be expressed as:

$$M_w = \left[-\frac{\ln(1 - a_w)}{C} \right]^{1/n} \tag{5.96}$$

where C and n are constants.

According to this model, a plot of $\ln(-\ln(1 - a_w))$ versus $\ln M_w$ should give a straight line. However, Rockland (1969) observed three localized isotherms that did not provide precise information on the physical state of water.

(g) Iglesias-Chirife Equation. Iglesias-Chirife (1978) proposed the following empirical equation:

$$\ln \left[M_w + \left(M_w^2 + M_{w,0.5} \right)^{1/2} \right] = C_1 a_w + C_2 \tag{5.97}$$

where $M_{w,0.5}$ is the moisture content at water activity of 0.5 and C_1 and C_2 are constants. This model was found to be suitable for foods with high sugar content such as fruits.

(h) The Guggenheim-Anderson-de Boer (GAB) Model. The term GAB model is based on the names Guggenheim (1966), Anderson (1946), and de Boer (1953), who derived the equation independently (Van den Berg, 1984).

The Guggenheim-Anderson-de Boer (GAB) model is suitable for many food materials over a wide range of water activity. It has many advantages over the others, such as having a viable theoretical background since it is refinement of Langmuir and BET theories of physical adsorption, showing a good description of nearly all food isotherms from zero to 0.9 water activity, being mathematically simple, and being easy to interpret since parameters have a physical meaning (Van den Berg, 1984).

The GAB model is expressed as:

$$M_w = \frac{M_0 C K a_w}{(1 - K a_w)(1 - K a_w + C K a_w)} \tag{5.98}$$

where M_0 is monolayer moisture content, and C and K are the adsorption constants which are related to the energies of interaction between the first and further sorbed molecules at the individual sorption sites. They can be theoretically expressed as:

$$C = c_0 \exp\left(\frac{\bar{H}_0 - \bar{H}_n}{RT} \right) \tag{5.99}$$

$$K = k_0 \exp\left(\frac{\bar{H}_n - \bar{H}_\ell}{RT} \right) \tag{5.100}$$

where c_0 and k_0 are the entropic accommodation factors; $\bar{H}_0$, $\bar{H}_n$, and $\bar{H}_\ell$ are the molar sorption enthalpies of the monolayer, of the multilayers on top of monolayer, and of the bulk liquid, respectively. R is the ideal gas constant and T is the absolute temperature, respectively.

Note that when K is 1, the GAB model becomes the BET equation. The BET model had been commonly used in water sorption of foods although it has limited range of water activity (0.05–0.45). In more recent years, the GAB sorption isotherm equation has been widely used in foods to represent the experimental data in the range of water activity of 0.10 to 0.90 which covers the majority of food products (Timmermann, Chirife, & Iglesias, 2001). Although BET and GAB models are closely related and follow the same statistical model, various researchers showed that monolayer capacity by BET is less while the energy constant by BET is greater than the GAB value. Because of this, there is a dilemma about which model describes the physical system better (Timmermann et al., 2001). In their recent work, Timmermann et al. (2001) solves this dilemma about the differences of constants obtained by BET and GAB models.

(i) Peleg Model. The Peleg (1993) model shows the same or better fit than the GAB model. It is a four-parameter model described as:

$$M_w = C_1 a_w^{C_3} + C_2 a_w^{C_4} \tag{5.101}$$

where C_1, C_2, C_3, and C_4 are constants and the constants $C_3 < 1$ and $C_4 > 1$.

It has been discussed that moisture sorption isotherms of foods can be described by more than one sorption model (Lomauro, Bakshi, & Labuza, 1985). To select the most suitable sorption model, degree of fit to the experimental data and simplicity of the model should be considered. Lomauro et al. (1985) showed that in more than 50% of fruits, meats, and vegetables analyzed, the GAB equation gave the best fit. Moisture sorption isotherm data obtained for cowpea, powdered cowpea, and protein isolate of cowpea at 10, 20, and 30°C were found to be excellently represented by the GAB equation (Ayranci & Duman, 2005). The sorption isotherm of dehydrated tomato slices in the water activity range of 0.08 to 0.85 was determined and the GAB model was found to be the best applicable model (Akanbi, Adeyemi, & Ojo, 2005). Tarigan, Prateepchaikul, Yamsaengsung, Srichote, and Tekasul (2005) found that the GAB equation gave the best fit to candle nuts within temperatures of 30, 40, 50, and 60°C. The adsorption and desorption isotherms of Gaziantep cheese were best described by GAB and quadratic polynomial equations (Kaya & Oner, 1995). Moisture sorption isotherms of grapes, apricots, apples, and potatoes were determined at 30, 45, and 60°C and the Halsey model equation gave the best fit for all fruits within the water activities and temperature ranges studied (Kaymak-Ertekin & Gedik, 2004). The Halsey and Peleg models adequately described the moisture sorption isotherms of pestil, which is Turkish grape leather at a water activity range of 0.06 to 0.98 (Kaya & Kahyaoglu, 2005). The sorption behavior of sweet potatoes was best described by Halsey equation (Fasina, 2005). The sorption behavior of mushrooms was explained by the BET model for a water activity range of 0.11 to 0.43 (Sahbaz, Palazoglu, & Uzman, 1999).

Example 5.5. Adsorption and desorption isotherms of grapes at 30°C were obtained by using saturated salt solutions with different relative humidity values. The data are given in Table E.5.5.1.

   (a) Draw the equilibrium moisture content curves for the adsorption and desorption processes. What is the type of sorption isotherm? Is hysteresis observed for sliced grapes?
   (b) Different isotherm models were explored for their fitting to the experimental data by nonlinear regression and a quite high coefficient of determination was found between the experimental

Table E.5.5.1 Equilibrium Moisture Content of Grapes During Adsorption and Desorption at 30°C[a]

Relative Humidity	Equilibrium Moisture Content (kg/100 kg dry solid)	
	Adsorption	Desorption
0	0	0
0.113	9.30	12.09
0.216	12.50	16.04
0.324	16.30	19.64
0.432	20.69	24.53
0.514	24.50	29.50
0.560	27.57	34.10
0.691	36.25	44.19
0.751	43.20	53.12
0.836	61.38	71.37
0.9	112.57	120.44

[a]From Kaymak-Ertekin, F., & Gedik, A. Sorption isotherms and isosteric heat of sorption for grapes, apricots, apples and potatoes. *Lebensmittel Wissenschaft und Technologie, 37,* 429–438. Copyright © (2004) with permission from Elsevier.

data and the values obtained by the Halsey model modified by Iglesias and Chirife (1976). From the nonlinear regression analysis, both C and n values for the Halsey model were determined. If the estimated value of n is given as 1.31, then what is the value of the constant C?

(c) If 200 g of sliced grapes with moisture content of 77% (w/w, wet basis) are dried by using circulating air with 40% relative humidity, what will be the final weight of the grapes?

Solution:

(a) Sliced grapes exhibit a type II isotherm. Hysteresis is observed (Fig. E.5.5.1).

(b) Using the Halsey model:

$$M_w = \left(-\frac{C}{\ln a_w} \right)^{1/n}$$

Arranging the equation as $M_w^n = \left(-\frac{C}{\ln a_w} \right)$

and plotting M_w^n versus $\left(-\frac{1}{\ln a_w} \right)$, the constant C is found from slope as 46.95 (Fig. E.5.5.2).

(c) First the amount of dry solid in grape is calculated:

$$(200 \text{ g})(1 - 0.77) = 46 \text{ g of dry solid}$$

40% relative humidity is equal to a_w of 0.4. From Fig. E.5.5.1 for a_w of 0.4, equilibrium moisture content is 24 g of moisture/100 g of dry solid.
The amount of moisture in dried grape is:

$$\left(\frac{24 \text{ g moisture}}{100 \text{ g dry solid}} \right) (46 \text{ g dry solid}) = 11 \text{ g}$$

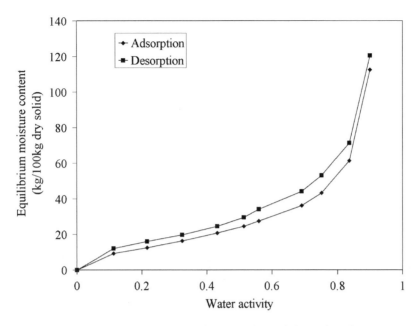

Figure E.5.5.1.   Equilibrium moisture content curve for adsorption and desorption of grapes.

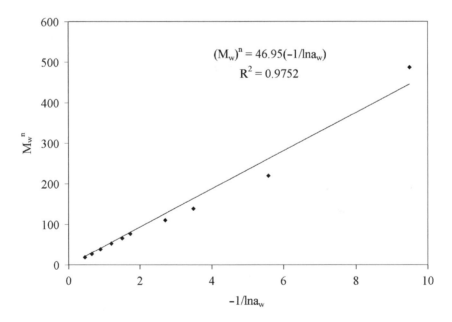

Figure E.5.5.2.   A plot of M_w^n versus $\left(-\dfrac{1}{\ln a_w} \right)$.

Then, the final weight of grapes is:

46 g dry solid + 11 g moisture = 57 g

Example 5.6. A sandwich is prepared using 60 g of cheese and 110 g of bread and then placed in a sealed container. The initial moisture content of the cheese and bread are 47% (wb) and 42% (wb), respectively. The sorption data given in Table E.5.6.1 are available.

(a) Calculate the equilibrium moisture content of the sandwich.
(b) Draw the sorption isotherms of cheese, bread, and sandwich (composite isotherm).

Solution:

(a) The initial moisture contents of cheese and bread in dry basis are calculated first:

$$m_{cheese} = \left(\frac{47}{100 - 47} \right) 100 = 88.68 \% \, (db)$$

$$m_{bread} = \left(\frac{42}{100 - 42} \right) 100 = 72.41 \% \, (db)$$

The moisture content at equilibrium is:

$$m_{eqlm} = f_{cheese} m_{cheese} + f_{bread} m_{bread}$$

where

$$f_{cheese} = \frac{\text{mass of solids in cheese}}{\text{mass of total solids in sandwich}} = \frac{60(100 - 47)}{60(100 - 47) + 110(100 - 42)} = 0.33$$

Table E.5.6.1. Moisture Content Data of Cheese and Bread at Different Water Activities

Water Activity	Moisture Content (%wb)	
	Cheese	Bread
0.331	10	3.29
0.432	11	4.63
0.544	13	6.71
0.699	18	10.93
0.755	23	12.90
0.813	27	15.18
0.851	30	16.79
0.946	40	20.94
0.976	49	24.17

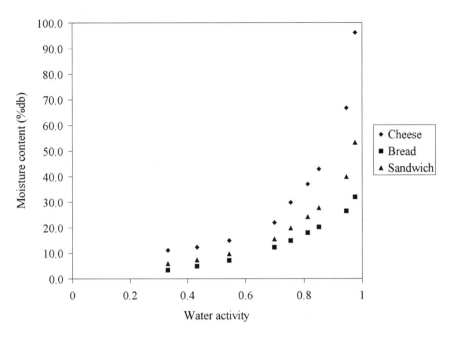

Figure E.5.6.1   Sorption isotherms of cheese, bread, and sandwich.

and
$$f_{\text{bread}} = \frac{\text{mass of solids in bread}}{\text{mass of total solids in sandwich}} = \frac{110(100 - 42)}{60(100 - 47) + 110(100 - 42)} = 0.67$$

$$\Rightarrow m_{\text{eqlm}} = 0.33 \times 88.68 + 0.67 \times 72.41$$

$$m_{\text{eqlm}} = 77.82 \ \%(\text{db})$$

(b) To prepare the sorption isotherms of cheese and bread, first their moisture content data given in wet basis are converted into dry basis (Table E.5.6.2). Then, the weighted averages of the moisture contents at each water activity are calculated as in the calculation of equilibrium moisture content in part (a) (Table E.5.6.2). The ratio of mass of cheese to bread is constant since the only change takes place in moisture content. Therefore, at each water activity f_{cheese} and f_{bread} are the same. Sorption isotherms of cheese, bread, and sandwich are shown in Fig. E.5.6.1.

PROBLEMS

5.1. Determine the boiling point temperature of 20% sucrose solution at 1 atm.
The molecular weight of sucrose is 342 g/mole and K_B is 0.51°C/m.

5.2. Table P.5.2.1 shows the equilibrium moisture content data of peppers in adsorption at 30°C.
(a) Draw the sorption isotherm for pepper.
(b) Different isotherm models were explored for their fitting to the experimental data by nonlinear regression and high coefficient of determination values were found between the experimental data and the values obtained by the Halsey model modified by Iglesias and

Table E.5.6.2. Weight Average Calculation and Isotherm Data

Water Activity	Moisture Content (%db)		
	Cheese	Bread	Sandwich
0.331	11.11	3.40	5.97
0.432	12.36	4.85	7.35
0.544	14.94	7.19	9.77
0.699	21.95	12.27	15.49
0.755	29.87	14.81	19.82
0.813	36.99	17.90	24.25
0.851	42.86	20.18	27.72
0.946	66.67	26.49	39.85
0.976	96.08	31.87	53.23

Chirife (1976). From the nonlinear regression analysis, both C and n values for the Halsey Model were determined. If the estimated value of n is given as 1.2, then what is the value of the constant C?

5.3. Sorption isotherm of cake was prepared at 40°C using the saturated salt solutions. Equilibrium moisture content data are given in Table P.5.3.1.

(a) Draw the sorption isotherm of cake.

(b) If your purpose is to obtain cake having 18% moisture in a dry basis, what should be the relative humidity of air during drying?

5.4. A solution is prepared by adding 1 g of sugar mannitol (MW 182.2 g/g-mole) in 100 g of water at 20°C.

(a) Assuming that mannitol and water form an ideal solution, what is the equilibrium vapor pressure of solution? The vapor pressure of pure water is 17.54 mmHg at 20°C.

Table P.5.2.1. Equilibrium Moisture Content of Peppers

Water Activity	Equilibrium Moisture Content (% db)
0.113	4.3
0.216	6.2
0.324	8.0
0.432	10.5
0.514	12.8
0.560	15.4
0.691	22.0
0.751	26.1
0.836	34.2
0.900	44.9

Table P.5.3.1. Equilibrium Moisture
Content Data of Cake Obtained at 40°C

Salts	Equilibrium Moisture Content (% db)
Lithium chloride	10
Magnesium chloride	15
Magnesium nitrate	21
Potassium iodide	32
Sodium chloride	43
Ammonium sulfate	51
Potassium chloride	57
Potassium nitrate	64
Potassium sulfate	72

(b) Calculate the activity of water as a solvent in this solution.

(c) Calculate the osmotic pressure of solution in mm Hg, by assuming that volume change due to addition of mannitol is negligible.

5.5. A certain biological solution has a freezing point of –0.3°C. If the solution is very dilute, determine its osmotic pressure at 25°C.

Data: $R = 0.082$ m^3·atm/kg-mole·K, $K_f = 1.86$

REFERENCES

Akanbi, C.T., Adeyemi, R.S., & Ojo, A. (2006). Drying characteristics and sorption isotherms of tomato slices. *Journal of Food Engineering, 28*, 45–54.

Anderson, R.B. (1946). Modification of the B.E.T. equation. *Journal of American Chemical Society, 68*, 686–691.

Ayranci, E., & Duman, O. (2005). Moisture sorption isotherms of cowpea (*Vigna unguiculata* L. Walp) and its protein isolate at 10, 20 and 30 °C. *Journal of Food Engineering, 70*, 83–91.

Bell, L.N., & Labuza, T.P. (2000). *Moisture Sorption: Practical Aspects of Isotherm Measurement and Use.* St Paul, MN: American Association of Cereal Chemists.

Brunauer, S., Emmett, P.H., & Teller, E. (1938). Adsorption of gases in multimolecular layers. *Journal of the American Chemical Society, 60*, 309–319.

Caurie, M.A. (1985). Corrected Ross equation. *Journal of Food Science, 50*, 1445–1447.

Cazier, J.B., & Gekas, V. (2001). Water activity and its prediction: A review. *International Journal of Food Properties, 4*, 35–43.

Crapiste, G.H., & Rotstein, E. (1982). Prediction of sorptional equilibrium data for starch-containing foodstuff. *Journal of Food Science, 47*, 1501–1507.

de Boer, J.H. (1953). *The Dynamical Character of Adsorption.* Oxford: Clarendon Press.

Fasina, O.O. (2005). Thermodynamic properties of sweetpotato. *Journal of Food Engineering* (http:/dx.doi.org/10.1016/ j.jfoodeng.2005.04.004).

Fortes, M., & Okos, MR. (1980). Drying theories: Their bases and limitations as applied to foods and grains. In A.S. Mujumdar (Ed.), *Advances in Drying*, Vol. 1 (pp. 119–154). Washington, DC: Hemisphere.

Greenspan L. (1977). Humidity fixed points of binary saturated aqueous solutions. *Journal of Research of the National Bureau of Standards Section A: Physics and Chemistry, 81*(1), 89–96.

Grover, D.W., & Nicol, J.M. (1940). The vapour pressure of glycerine solutions at 20°C. *Journal of the Society of Chemical Industry, 59*, 175–177.

Guggenheim, E.A. (1966). *Applications of Statistical Mechanics.* Oxford: Clarendon Press.

Halsey, G. (1948). Physical adsorption on non-uniform surfaces. *Journal of Chemical Physics, 16*, 931–937.

Henderson, S.M. (1952). A basic concept of equilibrium moisture. *Agricultural Engineering, 33*, 29–32.

Iglesias, H.A., & Chirife, J. (1976). A model for describing the water sorption behavior of foods. *Journal of Food Science, 41*, 984–992.

Iglesias, H.A., & Chirife, J. (1978). An empirical equation for fitting water sorption isotherms of fruits and related products. *Canadian Institute of Food Science of Technology Journal, 11*, 12–15.

Kapsalis, J.G. (1987). Influences of hysterisis and temperature on moisture sorption isotherms. In L.B. Rockland & L.R. Buchat (Eds.), *Water Activity: Theory and Applications to Food* (pp. 173–213). New York: Marcel Dekker.

Kaya, S., & Kahyaoglu T. (2005). Thermodynamic properties and sorption equilibrium of pestil (grape leather). *Journal of Food Engineering, 71*, 200–207.

Kaya, S., & Oner, M.D. (1995). Water activity and moisture sorption isotherms of Gaziantep cheese. *Journal of Food Quality, 19*, 121–132.

Kaymak-Ertekin, F., & Gedik, A. (2004). Sorption isotherms and isosteric heat of sorption for grapes, apricots, apples and potatoes. *Lebensmittel Wissenschaft und Technologie, 37*, 429–438.

Labuza T. P. (1984). *Moisture Sorption: Practical Aspects of Isotherm Measurement and Use.* St. Paul, MN: American Association of Cereal Chemists.

Labuza, T.P., Mizrahi, S., & Karel, M. (1972). Mathematical models for optimization of flexible film packaging of foods for storage. *Transactions of ASAE, 15*, 150–155.

Langmuir, I. (1918). The adsorption of gases on plane surfaces of glass, mica and platinum. *Journal of the American Chemical Society, 40*, 1361–1402.

Lewicki, P.P., & Pomaranska-Lazuka, W. (2003). Errors in static desiccator method of water sorption isotherm estimation. *International Journal of Food Properties, 6*, 557–563.

Lomauro, C.J., Bakshi, A.S., & Labuza, T.P. (1985). Evaluation of food moisture sorption isotherm equations.1. Fruit, vegetable and meat-products. *Lebensmittel Wissenschaft und Technologie, 18*(2), 111–117.

Nunes, R.V., Urbicain, M.J., & Rotstein, E. (1985). Improving accuracy and precision of water activity measurements with a vapor pressure manometer, *Journal of Food Science, 50*, 148–149.

Okos M.R., Narsimhan G., Singh R.K. & Weitnauer, A.C. (1992). Food Dehydration. In R. Heldman & D.B. Lund (Eds.), *Handbook of Food Engineering* (pp. 437–562). New York: Marcel Dekker.

Oswin, C.R. (1946). The kinetics of package life. III. The isotherm. *Journal of Chemical Industry (London)*, 65, 419–423.

Peleg, M. (1993). Assesment of a semi-empirical four parameter general model for sigmoid moisture sorption isotherm. *Journal of Food Process Engineering, 16*, 21.

Perry, R.H., & Chilton, C.H. (1973). *Chemical Engineering Handbook*, 5th ed. New York: McGraw-Hill.

Rahman, M.S. (1995). *Food Properties Handbook.* New York: CRC Press.

Rizvi, S.S.H. (2005). Thermodynamic properties of foods in dehydration.In M.A. Rao, S.S.H. Rizvi & A.K. Datta (Eds.), *Engineering Properties of Foods,* 3rd ed. (pp. 239–326) Boca Raton, FL: CRC Press Taylor & Francis.

Rockland, L.B. (1969). The practical approach to better low moisture foods: Water activity and storage stability. *Food Technology, 23*, 1241.

Ross, K.D. (1975). Estimation of water activity in intermediate moisture foods. *Food Technolology, 39*, 26–34.

Ruegg, M. (1980). Calculation of the activity of water in sulfuric acid solutions at various temperatures. *Lebensmittel Wissenschaft und Technologie, 13*, 22–24.

Sahbaz, F., Palazoglu, T.K., & Uzman, D. (1999). Moisture sorption and the applicability of the Brunauer-Emmet-Teller (BET) equation for blanched and unblanched mushroom. *Nahrung, 43*, 325–329.

Smith, S.E. (1947). The sorption of water vapor by high polymers. *Journal of the American Chemical Society, 69*, 646–651.

Spiess, W.E.L., & Wolf, W. (1987). Critical evaluation of methods to determine moisture sorption isotherms. In L.B. Rockland and L.R. Beuchat (Eds.), *Water Activity: Theory and Applications to Food* (pp. 215–233). New York: Marcel Dekker.

Tarigan, E., Prateepchaikul, G., Yamsaengsung, R., Srichote, A., & Tekasul, P. (2005). Sorption isotherms of shelled and unshelled kernels of candle nuts. *Journal of Food Engineering* (http://dx.doi.org/10.1016/j.jfoodeng.2005.04.030).

Timmermann, E.O., Chirife, J., & Iglesias, H.A. (2001). Water sorption isotherms of foods and food stuffs: BET or GAB parameters? *Journal of Food Engineering, 48*, 19–31.

Troller, J.A. (1983). Methods to measure water activity. *Journal of Food Protection, 46*, 129–134.

Troller, J.A., & Christian, J.H.B. (1978). *Water Activity and Food.* New York: Academic Press.

Van den Berg, C. (1984). Description of water activity of foods for engineering purposes by means of the GAB model of sorption, In B.M. McKenna (Ed.), *Engineering and Foods* (pp. 311–321). London: Elsevier.

CHAPTER 6

Surface Properties of Foods

SUMMARY

In this chapter, the principles and measurement methods of surface tension and interfacial tension are discussed. Information about colloidal systems in foods is also given.

Surface tension is defined as work required to extend a surface under isothermal conditions. If the surface forces take place at the interfaces, it is called interfacial tension. The methods for measuring surface tension are capillary rise, drop weight, bubble pressure, tensiometer, and dynamic methods. Substances that reduce the surface tension of a liquid are known as surface-active materials. Emulsifiers and hydrocolloids used in food systems are typical examples of surface-active materials. With certain modifications, the same methods used for surface tension can also be used for measuring interfacial tension.

Colloidal dispersion is the two-phase system in which the particles in the dispersed phase are between 1 and 1000 nm in diameter. Colloidal systems in foods can be categorized into four groups—sol, gel, emulsion, and foam—based on the states of matter in the continuous and dispersed phases. Proteins can act as both emulsifiers and stabilizers in food systems. The major difference between an emulsifier and a stabilizer is that the emulsifier is used to obtain short-term stabilization but stabilizers supply long-term emulsion stability.

6.1 SURFACE TENSION

You might have noticed water in spherical droplets on the surface of a leaf or emerging from tap. This can be explained by surface tension.

A molecule in the bulk of the liquid is attracted in all directions that cancel each other (Fig. 6.1). However, on the surface, the molecules are attracted across the surface and inward since the attraction of the underlying molecules is greater than the attraction of the vapor molecules on the other side of the surface. Therefore, the surface of the liquid is in a state of tension. This causes water to pull itself into a spherical shape which has the least surface area. Molecules at the surface of the liquid are attracted inward because of the van der Waals intermolecular attractions. This creates a force in the surface that tends to minimize the surface area and this force is known as surface tension. Surface tension can be defined as the tendency of a surface of a liquid to behave like a stretched elastic membrane. If the surface is stretched, the free energy of the system is increased.

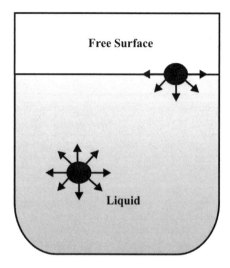

Figure 6.1 Interactions of water molecules inside and near the surface of liquid.

As can be seen in Fig. 6.2, the small droplet adapts its shape to an almost perfect sphere since it has the least surface area per unit volume. The shape becomes flatter as the size increases because of gravity since gravity is a function of unit volume whereas surface tension is a function of surface area. Therefore, gravity is more important for particles having larger sizes.

You can observe the effects of surface tension by conducting a simple experiment at home. When you shake black pepper into a glass of water, you will observe that the pepper will float because of surface tension. When a drop of soap or detergent is added to the water, the pepper will sink. Soap or detergents have the ability to reduce the surface tension of liquids.

The surface tension (σ) is expressed as free energy per unit surface area or work required to extend a surface under isothermal conditions. It is also defined as the force per unit length on the surface that opposes the expansion of the surface. This definition can be seen in Fig. 6.3, in which a bar is pulled with force F to expand a liquid film that is stretched like a bubble film on a wire frame.

$$\sigma = \frac{\text{Work done}}{\text{Increase in area}} = \frac{FL}{2Ld} = \frac{F}{2d} \tag{6.1}$$

where d is the distance between wires A and B, L is the distance that the bar is advanced to the left, and the factor 2 is introduced since there are two liquid surfaces (one at the front and one at the back).

The surface tension has dimensions of force per unit length. In the SI system, the unit for surface tension is N/m. The surface tension values of some liquid foods are given in Table 6.1.

Figure 6.2 Effects of surface tension on droplets having different sizes.

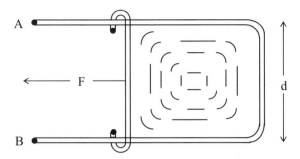

Figure 6.3   Schematic diagram for determination of the surface tension of a liquid.

Water has a very high surface tension value (Table 6.1). Liquids that have high surface tension values also have high latent heat values. The surface tensions of most liquids decrease as the temperature increases. The surface tension value becomes very low in the region of critical temperature as the intermolecular cohesive forces approach zero. The surface tensions of liquid metals are large in comparison with those of organic liquids. For example, the surface tension of mercury at 20°C is 435.5 mN/m (Weast, 1982). As can be seen in Table 6.1, alcohol in wine reduces the surface tension.

When dealing with surface properties, several equations give us some insight into the physical phenomena occurring during the processes related to surface rather than volume. They are the Laplace, Kelvin, Young, and Gibbs adsorption equations.

Table 6.1   Surface Tensions of Some Liquids at 20°C

Liquid	Surface Tension (mN/m)
Water[a]	72.75
Milk[b]	42.3–52.1
Skim milk (0.04% milk)[c]	52.7
Cream (34% milk)[c]	45.5
Cotton seed oil[d]	35.4
Coconut oil[d]	33.4
Olive oil[e]	33.0
Sunflower oil[f]	33.5
Wine (Chardonnay) (10.8% ethanol)[g]	46.9
Diluted wine (Chardonnay) (2.7% ethanol)[g]	60.9

[a]From data of Weast (1982).
[b]From data of Jenness, Shipe, and Sherbon (1974).
[c]From data of Witnah (1959).
[d]From data of Flingoh and Chong Chiew(1992).
[e]From data of Powrie and Tung (1976).
[f]From data of Ould-Eleya and Hardy (1993).
[g]From data of Peron et al. (2000).

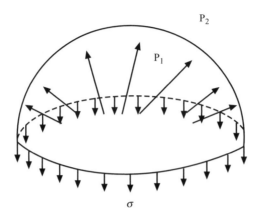

Figure 6.4 Half shape of a liquid droplet in vapor.

6.2 LAPLACE EQUATION

Consider a liquid droplet in spherical shape in vapor (Fig. 6.4). The internal pressure of the liquid is at pressure P_1 while the environment is at P_2. The surface tension of the liquid tends to cause the bubble to contract while the internal pressure P_1 tries to blow apart. Therefore, two different forces need to be considered.

Force due to a pressure difference between the pressure inside the droplet and the vapor outside is:

$$F = (P_1 - P_2)\pi r^2 \tag{6.2}$$

Force due to surface tension is:

$$F = 2\pi r \sigma \tag{6.3}$$

At equilibrium, these two forces are balanced. Then:

$$\Delta P = P_1 - P_2 = \frac{2\sigma}{r} \tag{6.4}$$

Equation (6.4) is known as the Laplace equation. This equation can be used to determine the relationship between the surface tension and the rise or depression of a liquid in a capillary as shown in Fig. 6.5. When a capillary tube is immersed in a liquid vertically, the liquid rises or depresses in the tube. The rise or depression of the liquid depends on the contact angle that liquid makes with the wall. The contact angle is the angle between the tangent to the surface of a liquid at the point of contact with a surface and the surface itself. If the contact angle is less than 90°, liquid will rise whereas it will depress for a contact angle greater than 90°.

If the capillary is of very small diameter, the meniscus will be in the shape of a sphere and the radius of curvature of liquid surface in the capillary can be expressed as:

$$r = \frac{r_t}{\cos \theta} \tag{6.5}$$

where r_t is the capillary tube radius and θ is the contact angle.

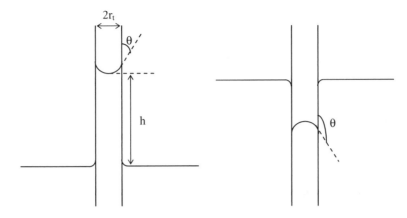

Figure 6.5   Rise or depression of a liquid in a capillary.

The difference between pressure in the liquid at the curved surface and pressure at the flat surface is:

$$\Delta P = \frac{2\sigma \cos \theta}{r_t} \tag{6.6}$$

Atmospheric pressure pushes the liquid up until pressure difference between liquid at the curved surface and liquid at the plane surface is balanced by the hydrostatic pressure:

$$\frac{2\sigma \cos \theta}{r_t} = hg(\rho_l - \rho_v) \tag{6.7}$$

where ρ_l is the density of the liquid, ρ_v is the density of vapor, and h is the height of the column.

Since the density of vapor is very small compared to that of the liquid, the equation can be approximated as:

$$\sigma = \frac{h \, g \, \rho_l \, r_t}{2 \cos \theta} \tag{6.8}$$

Example 6.1. Calculate the height of rise of water in a clean capillary tube of radius 0.001 cm if the density of water is 997 kg/m^3, surface tension is 73 dynes/cm, and the contact angle of water to the glass is 10°.

Solution:

$$\sigma = \left(73\frac{\text{dyne}}{\text{cm}}\right)\left(\frac{10^{-5}N}{1\,\text{dyne}}\right)\left(\frac{100\,\text{cm}}{1\,\text{m}}\right) = 7.3 \times 10^{-2}\,\text{N/m}$$

From Eq. (6.8):

$$\sigma = \frac{h \, g \, \rho_l \, r_t}{2 \cos \theta} \tag{6.8}$$

$$\Rightarrow h = \frac{2\cos(10°)\left(7.3 \times 10^{-2} \frac{N}{m}\right)}{\left(9.81 \frac{m}{s^2}\right)\left(997 \frac{kg}{m^3}\right)\left(10^{-5} m\right)} = 1.47\,m$$

6.3 KELVIN EQUATION

In addition to capillarity, another important consequence of pressure associated with surface curvature is the effect it has on thermodynamic activity. For liquids, the activity is measured by the vapor pressure. In another words, curvature of a surface affects the vapor pressure. This is expressed by the relationship of differential change in the area of the interface and the differential change in Gibbs free energy of the system. Starting from the first law of thermodynamics:

$$dU = dQ + dW \tag{6.9}$$

For a system in equilibrium:

$$dQ = TdS \tag{6.10}$$

and

$$dW = -PdV + \sigma dA \tag{6.11}$$

Combining the first and second laws of thermodynamics for an open-component system:

$$dU = TdS - PdV + \sigma dA + \mu dn \tag{6.12}$$

In the last term in Eq. (6.12), μ is the chemical potential and dn is the change in the number of moles in the system. This term gives the change in Gibbs free energy when dn moles of a substance are added to the droplet without changing the area of the surface, which is the case for a planar surface.

The Gibbs free energy of the system can be expressed by:

$$G = H - TS \tag{6.13}$$

Substituting $H = U + PV$:

$$G = U + PV - TS \tag{6.14}$$

Differentiating Eq. (6.14):

$$dG = dU + PdV + VdP - TdS - SdT \tag{6.15}$$

Inserting Eq. (6.12) for dU:

$$dG = \sigma dA + \mu dn - SdT + VdP \tag{6.16}$$

For a sphere, volume change is $dV = 4\pi r^2 dr$ and area change is $dA = 8\pi rdr$. Thus:

$$dA = \frac{2dV}{r} = \frac{2\overline{V}dn}{r} \tag{6.17}$$

where $\overline{V}$ is the molar volume of the liquid. Combining Eq. (6.17) with (6.16), Eq. (6.18) is obtained:

$$dG = \left(\frac{2\overline{V}\sigma}{r} + \mu_{\text{planar}}\right)dn - SdT + VdP \tag{6.18}$$

The chemical potential of the liquid in the droplet is given by:

$$\mu_{droplet} = \frac{2\overline{V}\sigma}{r} + \mu_{planar} \tag{6.19}$$

If the vapor behaves like a perfect gas:

$$\mu_{droplet} = \mu_{planar} + RT \ln \frac{P}{P^0} \tag{6.20}$$

where P^0 is the vapor pressure of the planar surface. Combining Eq. (6.20) with Eq. (6.19), the Kelvin equation is obtained:

$$\ln \left(\frac{P}{P^0} \right) = \frac{2\overline{V}\sigma}{rRT} \tag{6.21}$$

This equation gives the vapor pressure (P) of a droplet having a radius r. The assumption to obtain the Kelvin equation is that the surface tension is independent of the curvature of the surface. If the sign of one part of the equation changes, it results in vapor pressure of the concave surface of a liquid in the capillary or a small bubble.

Example 6.2. The vapor pressure of water is 3167.7 Pa at 25°C. Calculate the vapor pressure of water at 25°C in droplets of radius 10^{-8} m if the surface tension for water is 71.97 dyne/cm at 25°C. Density and molecular weight of water are 997 kg/m³ and 18 kg/kg-mole, respectively. The gas constant, R, is 8314.34 J/kg-mole K.

Solution:

$$\sigma = \left(71.97 \frac{dyne}{cm} \right) \left(\frac{10^{-5} N}{1 \, dyne} \right) \left(\frac{100 \, cm}{1 \, m} \right) = 7.2 \times 10^{-2} \, N/m$$

Molar volume is calculated by dividing molecular weight by density of sample:

$$\overline{V} = \frac{18 \, kg/kgmole}{997 \, kg/m^3} = 0.018 \, m^3/kgmole$$

Using the Kelvin equation (6.21):

$$\ln \left(\frac{P}{P^0} \right) = \frac{2\overline{V}\sigma}{rRT} \tag{6.21}$$

$$\ln \left(\frac{P}{3167.7} \right) = \frac{2(0.018)(7.2 \times 10^{-2})}{(10^{-8})(8314.34)(298)}$$

P is calculated as 3517 Pa.

6.4 SURFACE ACTIVITY

Substances that reduce the surface tension of a liquid at very low concentrations are called surface active. Emulsifiers and hydrocolloids are good examples for surface-active materials. These materials have both polar or hydrophilic and nonpolar or lipophilic groups. They orient themselves at the interface between the two phases. These substances concentrate at the surface and give large decreases in surface tension. This can be expressed quantitatively by the Gibbs adsorption isotherm.

$$\Gamma = -\frac{1}{mRT}\frac{d\sigma}{d\ln a} \cong -\frac{1}{RT}\frac{d\sigma}{d\ln c} \tag{6.22}$$

where Γ is adsorption (excess concentration) of solute at the surface (kg-mole/m^2), σ is surface tension (N/m), R is the gas constant (8314.34 J/kg-mole · K), a is activity of solute in the bulk solution, c is concentration of solute in the bulk solution (kg-mole/m^3), and m is a constant with a value of 1 for dissociating substances.

The sign of excess concentration (Γ) is opposite to the sign of the change of surface tension with concentration or activity of solute in solution.

Example 6.3. The surface tension of aqueous solution of butanol was measured at 20°C and the results were given in Table E.6.3.1. Use the Gibbs isotherm to find the surface excess at $c = 0.1$ mol/L.

Solution:

Using Gibbs adsorption isotherm (Eq. 6.22), σ is plotted with respect to $\ln c$ (Fig. E.6.3.1).

$$\Gamma \cong -\frac{1}{RT}\frac{d\sigma}{d\ln c} \tag{6.22}$$

The slope of Fig. E.6.3.1 at $c = 0.1$ mol/L will give the surface excess at that point.
The polynomial equation of the curve is:

$$\sigma = -6.145(\ln c)^2 - 45.143(\ln c) - 15.253$$

$$\frac{d\sigma}{d(\ln c)} = -12.29(\ln c) - 45.143$$

at $c = 0.1$ or at $\ln c = -2.302$:

$$\frac{d\sigma}{d(\ln c)} \text{ is found as } - 16.85 \text{ mN/m.}$$

Table E.6.3.1. The Surface Tension Values of Aqueous Butanol Solutions at Different Concentrations

c (mol/L)	σ (mN/m)
0.0264	68.17
0.0536	62.78
0.1050	56.36
0.2110	40.12
0.4330	18.05

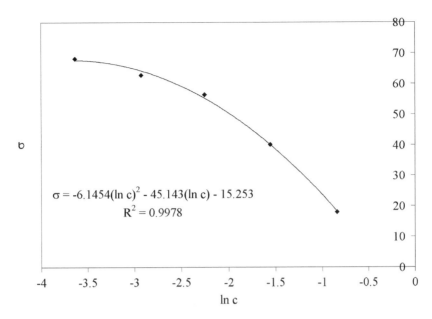

Figure E.6.3.1 Surface tension versus ln c graph.

Inserting the value of $\frac{d\sigma}{d(\ln c)}$ into Eq. (6.22), the excess concentration of solute at surface is determined as:

$$\Gamma = -\frac{1}{\left(8314.34 \, \frac{J}{kgmole \cdot K}\right)(293K)}\left(\frac{1J}{1N \cdot m}\right)\left(-16.85 \times 10^{-3} \, \frac{N}{m}\right) = 6.92 \times 10^{-9} \, \frac{kgmol}{m^2}$$

6.5 INTERFACIAL TENSION

Surface tension appears in situations involving either free surfaces (liquid–gas or liquid–solid boundaries) or interfaces (liquid–liquid boundaries). If it takes place in interfaces, it is called interfacial tension.

Interfacial tension arises at the boundary of two immiscible liquid due to the imbalance of intermolecular forces. Emulsifiers and detergents function by lowering the interfacial tension. Generally, the higher the interfacial tension, the lower is the solubility of the solvents in each other. Interfacial tension values between water and some organic solvents and oils are given in Table 6.2.

6.6 YOUNG AND DUPRE' EQUATIONS

In a liquid droplet on a solid surface, the interfacial tensions involving solid–liquid, solid–vapor, and liquid–vapor should be considered (Fig. 6.6). All the interfacial tensions are acting to minimize interfacial energy. The forces acting are balanced at equilibrium and Young's equation is

Table 6.2 Interfacial Tension Values Between Water and Some Materials

Material	Temperature ($^{\circ}C$)	Interfacial Tension (mN/m)
n-Hexane[a]	20	51.1
Peanut oil[b]	25	18.1
Olive oil[b]	25	17.6
Coconut oil[b]	25	12.8
Mercury[c]	20	375.0

[a]From data of Kaye and Laby (1973).
[b]From data of Powrie and Tung (1976).
[c]From data of Shaw (1970).

obtained:

$$\sigma_{sv} = \sigma_{lv} \cos \theta + \sigma_{sl} \tag{6.23}$$

where θ is the contact angle; σ is surface tension (N/m); and the subscripts s, l, and v denote solid, liquid, and vapor, respectively.

If θ is greater than 90°, the liquid does not wet the solid and tends to move about on the surface and does not enter capillary pores. A liquid completely wets the solid only when θ is zero (Adamson, 1990).

Young's equation relates solid and liquid surface tensions (work necessary to create one solid or liquid surface unity), solid–liquid interfacial tension (work necessary to create one interface unity between solid and liquid), and the contact angle.

An energy balance known as the Dupre' equation defines work of adhesion between solid and liquid. The Dupre' equation relates adhesive and adherent surface tensions and the work of adhesion (W_a) (McGuire, 2005):

$$W_a = \sigma_{sv} + \sigma_{lv} - \sigma_{sl} \tag{6.24}$$

The theory of thermodynamic adsorption was based on Young and Dupree energy equations (Michalski, Desobry, & Hardy, 1997). The thermodynamic adsorption theory was broadly developed by various researchers (Dann, 1970; Fowkes, 1964, 1972, 1983; Girifalco & Good, 1957; Li & Neumann, 1992: Van Oss, Good, & Busscher, 1990). These researchers showed that adhesion was

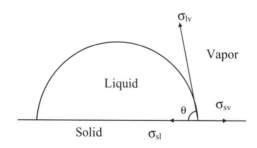

Figure 6.6 Forces involved in Young's equation.

due to electrodynamic intermolecular forces acting at the liquid–liquid, liquid–solid, and solid–solid interfaces. Interfacial attraction was then expressed in terms of reversible work of adhesion related to the surface tensions of materials. The details of mechanisms of adhesion can be found in the review of Michalski et al. (1997).

6.7   COLLOIDAL SYSTEMS IN FOODS

All colloidal systems have two phases: a continuous phase and a discontinuous or dispersed phase. Colloidal dispersion is a two-phase system in which the particles in the dispersed phase are between 1 and 1000 nm in diameter. For this reason, most manufactured foods can be considered as food colloids and many contain hydrocolloids that are added to control stability and rheological properties. Food hydrocolloids are high molecular weight hydrophilic biopolymers used in food products to control their texture, flavor, and shelf-life.

Colloidal systems in foods can be categorized into four groups based on the states of matter constituting the two phases, which are sols, gels, emulsions, and foams.

6.7.1   Sols

A sol can be defined as a colloidal dispersion in which a solid is the dispersed phase and liquid is the continuous phase. Gravy, stirred custard, and other thick sauces can be given as examples of sols. The proper ratio of the ingredients is necessary to achieve the desired viscosity of the sols at a certain temperature. If they are unacceptably thick, they can either be heated or more water can be added to reduce their viscosity.

Pectin is hydrophilic and attracts a layer of water that is bound tightly to the molecules by hydrogen bonds, like other carbohydrates and many proteins. Thus, water forms an insulating shield for the pectin or other hydrophilic colloid, providing a layer that inhibits bonding between the molecules of the colloidal substances (Fig. 6.7). Sols can be transformed into gels as a result of reduction in temperature. The solids in the discontinuous phase move with increasing difficulty through the continuous liquid phase and eventually start to associate with each other. In a pectin gel, the pectin molecules are the continuous phase and the liquid is the dispersed phase while in the pectin sol, the pectin molecules are the dispersed phase and the liquid is the continuous phase. Sols may be formed as a preliminary step in making a gel. Jams and jellies made with pectin are common examples that form a sol prior to the desired gel structure.

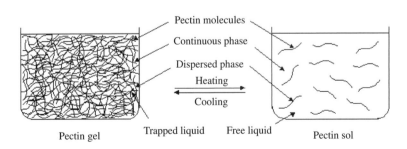

Figure 6.7   Pectin gel and sol.

6.7.2 Gels

A gel is the reverse of a sol in which a solid matrix is the continuous phase and a liquid is the discontinuous phase. The solid in the gel is sufficiently concentrated to provide the structure needed to prevent flow of the colloidal system.

Some of the free liquid may be released if the gel structure is cut. This phenomenon is known as syneresis. The type of solid and its concentration in the gel are important in determining the amount of syneresis. Syneresis may be undesired in some products such as jelly but may be useful in cheese production.

6.7.3 Emulsions

An emulsion is a colloidal system in which a liquid is dispersed as droplets in another liquid with which it is immiscible. Emulsions can be classified as oil-in-water (o/w) and water-in-oil emulsions (w/o). In oil-in-water emulsions, oil is dispersed in water as droplets. The typical example for o/w emulsions is mayonnaise. In water-in-oil emulsions, such as butter, droplets of water are dispersed in oil.

Viscosities of the emulsions are very high as compared to the viscosities of either of the liquids. For example, the viscosity of mayonnaise is higher than that of the vinegar and oil used in preparing mayonnaise. The stability of emulsions can be determined by the viscosity of the continuous phase, the presence and concentration of emulsifier, the size of the droplets, and the ratio of dispersed phase to the continuous phase.

When an emulsion is subjected to some stress such as freezing or centrifugation, it may break. That is, the two liquids separate into discrete phases.

In emulsion systems it is necessary to determine the difference between an emulsifier and a stabilizer (Dickinson, 1992). The emulsifiers aid emulsion formation and short-term stabilization by interfacial action. An emulsifier is a small molecule surfactant that is amphiphilic, having both polar or nonpolar parts, and thus has attraction toward both phases of the emulsion. The emulsifying agent collects around the surface of the dispersed spheres; as a result the droplets cannot touch each other directly and coalescence. Monoglycerides, polysorbates, sucrose esters, and lecithin are examples for emulsifiers. Egg yolk is the best common food naturally containing emulsifying agents. The primary emulsifying agent found in egg yolk is lecithin. Small molecules are not so effective for supplying long-term stability.

The amount of emulsifying agent present has a significant effect on the stability of the emulsion. Enough agent must be present to form a complete monomolecular layer around each droplet but there is no benefit of using more emulsifying agent than the required amount.

Stabilizers, on the other hand, are generally proteins or polysaccharides supplying long-term emulsion stability, possibly by an adsorption mechanism but not necessarily so. The main stabilizing action of food polysaccharides is by modifying viscosity or gelation in aqueous phase. Proteins act as both emulsifier and stabilizer since they have high tendency to adsorb at oil–water interfaces to form stabilizing layers around oil droplets (Dickinson, 2003).

An emulsifier should be surface active, meaning that it has an ability to reduce the surface tension at the oil–water interface. The lower the interfacial tension, the greater the extent to which droplets can be broken up during intense shear or turbulent flow (Walstra & Smulders, 1998).

An amphiphilic characteristic is required for a polymer to be surface-active. If it is a hydrocolloid, it must contain numerous hydrophobic groups. These groups will enable adsorbing molecules to adhere to and spread out at the interface which will protect the newly formed droplets. An ideal emulsifier is composed of species with relatively low molecular weight, with good solubility in aqueous continuous phase. For a biopolymer to be effective in stabilizing dispersed particles or emulsion droplets, it should have a strong adsorption characteristic, complete surface coverage, formation of a thick steric, and charged stabilizing layer (Dickinson, 2003). Strong adsorption implies that the amphiphilic polymer has a substantial degree of hydrophobic character to keep it permanently at the interface. Complete surface coverage means that the polymer is sufficient to fully saturate the surface. Formation of a thick steric stabilizing layer implies that the polymer is hydrophilic and has a high molecular weight (10^4–10^6 Da) within an aqueous medium with good solvent properties. Formation of a charged stabilizing layer implies the presence of charged groups on the polymer that contribute to the net repulsive electrostatic interaction between particle surfaces, especially at low ionic strength.

Proteins are surface-active compounds and comparable with low molecular weight surfactants (emulsifiers). They result in lowering of the interfacial tension of fluid interfaces. Proteins emulsify an oil phase in water and stabilize the emulsion. The increased resistance of oil droplets to coalescence is associated with protein adsorption at the interfacial surface and the formation of interfacial layer having high viscosity. In comparison to the low molecular weight surfactants, proteins have the ability to form an adsorption layer with high strength and increase the viscosity of the medium which results in high stability of emulsion (Garti & Leser, 2001). Low molecular weight surfactants (LMW) commonly used in foods are mono- and diglycerides, polysorbates, sorbitan monostearate, polyoxyethylene sorbitan monostearate, and sucrose esters. They lower the interfacial tension to a greater extent than high molecular weight surfactants such as proteins and gums (Table 6.3). LMW surfactants have higher adsorption energies per unit area than proteins and gums but proteins and gums can adsorb at the interface with several segments. Although LMW surfactants are more effective than the others in reducing interfacial tensions, emulsions formed by them are less stable. This is due to the fact that steric repulsion between the protein-covered oil droplets is very effective against aggregation. In addition, since proteins have higher molecular weight, their adsorption and desorption are slower than the adsorption and desorption of LMW surfactants (Bos & van Vliet, 2001).

Surfactants can be classified as nonionic or ionic according to the charge of the head group. They are also categorized based on their ability to dissolve in oil or water. A measure of this is expressed by hydrophilic/lipophilic balance (HLB). HLB number can be calculated by using the

Table 6.3 General Characteristics of Interfacial Properties of LMW Surfactants and Proteins[a]

Interfacial Properties	LMW Surfactants	Proteins
Number of molecules	10^{-5} mol/m^2	10^{-7} mol/m^2
Adsorbed amount	1.0–2.0 mg/m^2	2.0–3.0 mg/m^2
Adsorption	Reversible	Practically irreversible
Molecular size and shape	Cylindrical ($1 \times 1 \times 2$ nm)	Often globular (4–5 nm)
Conformational changes upon adsorption	No	Yes
Equilibrium interfacial tension at air-water interface	40–22 mN/m	57–47 mN/m
Value of surface tension gradient	High	Low

[a]From Bos, M. A., & van Vliet, T. Interfacial rheological properties of adsorbed protein layers and surfactants: A review. *Advances in Colloid and Interface Science, 91*, 437–471. Copyright © (2001) with permission from Elsevier.

Table 6.4 Ranges of HLB Values for Different Applications.[a]

HLB Value	Application
3–6	Emulsifiers of w/o emulsions
7–9	Wetting agents
8–18	Emulsifiers of o/w emulsions
13–15	Detergents
15–18	Solubilizers

[a]From Bos, M. A., & van Vliet, T. Interfacial rheological properties of adsorbed protein layers and surfactants: A review. *Advances in Colloid and Interface Science, 91*, 437–471. Copyright © (2001) with permission from Elsevier.

relationship:

$$HLB = 7 + \sum \text{hydrophilic group numbers} - \sum \text{lipophilic group numbers} \tag{6.25}$$

HLB value is the rate of the weight percentage of hydrophilic groups to the weight percentage of hydrophobic groups in the emulsifier molecule. Emulsifiers with HLB values below 9 are lipophilic, those with HLB values between 8 and 11 are intermediate, and those with HLB values between 11 and 20 are hydrophilic (Lewis, 1996). Table 6.4 shows the application of surfactants and their HLB values. Diacetyl tartaric acid ester of monoglycerides (DATEM), which is a common emulsifier used in baked products has a HLB value of 9.2 (Krog & Lauridsen, 1976). Sorbitan monostearate, with a HLB value of 5.7, can be used in cake mixes and cacao products.

Generally a combination of emulsifiers is necessary to achieve a stable emulsion. HLB value of a mixed emulsifier system $(HLB)_m$ containing emulsifier A and B can be calculated by using HLB values (HLB_A and HLB_B) and mass fraction (X_A^w and X_B^w) of individual emulsifiers:

$$(HLB)_m = X_A^w(HLB_A) + X_B^w(HLB_B) \tag{6.26}$$

Some food hydrocolloids are sufficiently surface active and act as emulsifying agents in oil-in-water emulsions. Their emulsifying properties come from the proteinaceous material covalently bound or physically associated with carbohydrate polymer. The emulsions produced are coarser than those with low molecular weight water soluble surfactants or milk proteins at the same emulsifier/oil ratio. After hydrocolloids are strongly adsorbed at the oil–water interface, hydrocolloids can be more effective than proteins or others. Polysaccharides affect emulsion stabilization by increasing the viscosity of the dispersing phase and surface adsorption. Surface tension values of several gums are given in Table 6.5.

Locust bean gum and guar gum were shown to decrease the surface tension of water at low concentrations (up to 0.5%) and surface tension of gum solutions was time dependent. Surface tensions decreased and adsorption rates increased significantly by increasing gum concentration (Garti & Leser, 2001).

Guar gum and locust bean gum are used for their thickening, water holding, and stabilizing properties. Although there are no hydrophobic groups in these gums, they function by modifying the rheological properties of aqueous phase.

Xanthan gum is used as a thickening agent in foods. Xanthan gum is used as a stabilizer for oil-in-water emulsions. It was proposed by some researchers that its stabilizing ability comes from the viscosity effect (Coia & Stauffer, 1987; Ikegami, Williams, & Phillips, 1992). The adsorption

Table 6.5 Surface Tension Values of Selected Gums[a]

Gum	Gum Weight (%)	σ (mN/m)
Tragacanth	0.6	42
Xanthan	0.6	43
Guar	0.7	55
Locust bean gum	0.7	50

[a]From Garti, N., & Leser, M.E. Emulsification Properties of Hydrocolloids. *Polymers for Advanced Technologies, 12*, 123–135. Copyright © (2001) with permission from John Wiley & Sons.

characteristic of xanthan gum was also studied. In general, the adsorption of gum onto liquid droplets needs different requirements than proteins. The steric stabilization of proteins on oil–water droplets cannot be observed in hydrocolloids since they lack the flexibility to reorient at the surface and hydrophobic moieties.

A possible adsorption mechanism of hydrophilic compounds onto the oil–water interphase is not well known. On the other hand, proteins containing hydrocolloids are good steric stabilizers because they contain a significant proportion of flexible hydrophobic groups to act as anchoring points and many hydrophilic groups to form the stabilizing chains. However, most polysaccharide polymers are sufficiently hydrophilic to be regarded as nonadsorbing.

6.7.4 Foams

Foams can be defined as a colloidal dispersion in which gas is the dispersed phase and liquid is the continuous phase. The dispersion medium is usually a liquid that is sometimes modified into a solid by heating or strengthened by a solid. Ice cream, whipped cream, or cake batter are some examples of foams. There is usually a third phase such as oil or fat in foam used in food systems. Foams are important because of their contribution to volume and texture of food products.

The inclusion of air reduces the density of the product. The term overrun is used in relation to foams to describe the amount of air incorporated. It is defined as the percentage increase in volume attributed to air and expressed as:

$$\text{Overrun} = \frac{V_f - V_l}{V_l} 100 \tag{6.27}$$

where V_f is the volume of the foam and V_l is the volume of the original liquid.

Liquids should have low surface tension to form foams. As a result of low surface tension, the liquid can be stretched or spread easily and does not coalescence. If the surface tension is high, formation of the foam will be difficult because of its resistance to spreading and strong tendency to expose the least surface area. If the liquid has high surface tension, the foam collapses quickly as a result of the tendency to coalescence.

Liquids should have low vapor pressure to prevent evaporation. This will help retain the gaseous phase within the liquid and minimize its tendency to break out into the surrounding medium.

Stability can be provided to foams if some solid matter can be incorporated into the films to increase the rigidity of the walls surrounding the gas. In protein foams, the denaturation of the protein provides stability to cell walls.

Table 6.6   The Relevance of the Surface and Bulk Properties to Bread Baking

Stage in Bread Baking Process	Surface Tension	Strain-Hardening	Resistance to Extension	Extensibility
Mixing	+	−	−	−
First proof	++	++	++	−
Final proof	−	++	+	+
Baking	?	++	+	++

++, very relevant; +, relevant; −, not relevant. From Kokelaar et al. (1994).

The interfacial properties of foam-based products are controlled by competition between surface-active agents that adsorb at the oil–water and air–water interfaces in these systems. Static (equilibrium) interfacial tension, dynamic interfacial properties, interfacial rheology, diffusion, and adsorption and desorption properties of the surfactants and proteins are important for the stability of foam-based foods (Xu, Nikolov, Wasan, Gonsalves, & Borwankar, 2003).

If a bubble in foam is considered, there are two forces acting on the bubble—dynamic forces and surface tension forces. The dynamic forces can be regarded as shear stress. The ratio of these forces, which is the Weber number, is used to determine the maximum stable bubble size possible (Walstra, 1983):

$$\text{We} = \tau \frac{D_{\text{max}}}{\sigma} \tag{6.28}$$

where τ is shear stress, σ is surface tension, and D is the diameter of the bubble.

During bread making in the mixing stage, air bubbles are entrapped in the gluten network. If We is greater than $\text{We}_{\text{critical}}$ the bubble breaks. The critical Weber number depends on the type of flow and the ratio of viscosity of gas in the bubble (gas) to that of continuous phase (dough). Lipids and added surfactants will lower the minimum air bubble radius formed during mixing which may lead to more and smaller bubbles in the dough resulting in a finer crumb structure. After mixing, the growth of large bubbles at the expense of smaller ones will occur, which is called disproportion. This can be retarded by the addition of surfactants. If the dough has enough strain-hardening properties, the gas bubbles that are large enough to expand at the existing carbon dioxide partial pressure will have a narrow bubble size distribution. This will lead to a regular crumb structure in the product (Kokelaar, Van Vliet, & Prins, 1994). The relevance of the surface and bulk properties to bread baking were studied and are summarized in Table 6.6.

6.8   MEASUREMENT OF CONTACT ANGLE AND SURFACE TENSION

6.8.1   Contact Angle Measurement Methods

Contact angle can be measured with a contact angle goniometer. This instrument consists of a light source, an illuminating stage on which the liquid drop–solid material three-phase system (liquid, vapor, solid) rests, and a telescope. By viewing the drop through the telescope, the contact angle is measured.

An inexpensive instrument of a video-based contact angle meters is also available to measure contact angle (McGuire, 2005). A detailed discussion about measurement of contact angle can be found in Michalski et al. (1997).

6.8.2 Surface Tension Measurement Methods

The methods for measuring surface tension are capillary rise, drop weight, bubble pressure, tensiometer, and dynamic methods. The same methods with certain modifications can be used for measuring interfacial tension.

The capillary rise method is a simple method. However, it is limited to the measurement of surface tension of pure liquids (Aveyard & Haydon, 1973). In the capillary rise method, a capillary tube is immersed in liquid vertically which then rises to a height h inside the tube. When equilibrium is reached, the upward forces of surface tension balance the downward forces due to the height of the column of liquid. Then:

$$\sigma = \frac{\rho_l \, g \, h \, r_t}{2 \cos \theta} \tag{6.8}$$

In the case of mercury, there may be a complete nonwetting of surface where angle of contact is $180°$ which leads $\cos \theta$ to be (-1). This causes capillary depression.

In the drop weight method, either the weight or the volume of the drop formed at the end of a capillary tube is measured. The droplet will grow to a size at which its weight overcomes the surface tension force. At this moment, the droplet will detach from the tube and its weight can be measured by collecting it in a small pan and weighing it. By equating the forces acting on the droplet, surface tension force, and gravitational force as the droplet breaks, surface tension can be determined as:

$$\sigma = \frac{mg}{2\pi \, r_t} \tag{6.29}$$

where m is the mass of the droplet and r_t is the external radius of the capillary tube.

This equation is an approximation since in practice the entire drop never falls completely from the tip. Therefore, a correction factor (F) is incorporated into Eq. (6.29) (James, 1967).

$$\sigma = \frac{Fmg}{r_t} \tag{6.30}$$

Problems of absolute determinations can also be overcome by using a fluid of known surface tension. If m_1 and m_2 are the drop masses of two fluids of surface tension σ_1 and σ_2, then the ratio can be written as:

$$\frac{\sigma_1}{\sigma_2} = \frac{m_1}{m_2} \tag{6.31}$$

There are some studies in literature in which the drop weight method was used to determine surface tension of whole and skim milk (Bertsch, 1983) and milk protein solutions (Arnebrant & Naylander, 1985).

For measuring surface tension of viscous and heterogeneous fluids such as batter, the bubble pressure method is used. In this method, a capillary tube is immersed into the sample (Fig. 6.8). Distilled water is then dropped into a vertical sigmoidal tube connected to a capillary to increase its internal pressure. A small bubble is formed at the end of capillary and floated to the surface of the product. When the radius of curvature is minimum, surface tension of the sample is calculated from the internal pressure

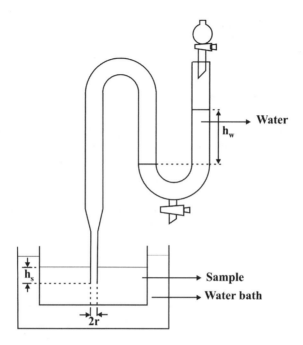

Figure 6.8 Bubble method. [From Sasaki,K., Shimaya, Y., Hatae, K., & Shimada, A. Surface tension of baked food batter measured by the maximum bubble pressure method. *Agricultural & Biological Chemistry, 55*, 1273–1279. Copyright © (1991) with permission from Japan Society for Bioscience, Biotechnology and Agrochemistry.]

of bubble by using the Laplace equation:

$$\rho_w g h_w - \rho_s g h_s = \frac{2\sigma}{r} \tag{6.32}$$

where h_w is the height difference between water surfaces in the sigmoidal tube, h_s is the length from the sample surface to the capillary, ρ_w is the water density, and ρ_s is the sample density.

Surface tensions of solutions containing surface-active compounds can be determined by the Du Nouy ring tensiometry method. In the literature, various studies are described in which this method was used to measure surface tension of oils (Flingoh & Chong Chiew, 1992), milks (Duthie & Jenson, 1959; Watson, 1956, 1958), aqueous solutions of hydroxymethylcellulose (HPMC) (Riedl, Szklenarik, Zelko, Marton, & Racz, 2000), and chitosan coating solution (Choi, Park, Ahn, Lee, & Lee, 2002).

In the Du Nouy method, a platinum ring is suspended above the liquid sample. It is dipped into the sample and pulled out. The force necessary to pull the ring through the surface is measured. The equation for the ring method is:

$$\sigma = \frac{mg}{2c} \tag{6.33}$$

where m is the mass and c is the circumference of the ring.

Equation (6.33) gives apparent surface tension and it is necessary to use a correction factor. The magnitude of the correction factor depends on the dimension of the ring and density of the liquid. The correction factors are supplied with the manufacturer of the instrument. This method gives static surface tension for fast-adsorbing substances. Otherwise, it gives dynamic surface tension.

The oscillating jet method is a dynamic method that is used to determine the surface tension of skim milk (Kubiak & Dejmek, 1993). This method is based on the mechanical instability in a jet of liquid sample emerging from an elliptical orifice. The circular shape of the drops tends to be restored by sample surface tension. However, as a result of the liquid inertia, oscillations around a circular cross-sectional shape occur. A parallel light beam applied perpendicular to the jet axis will then deviate on the drops. Then, sample surface tension can be calculated from oscillation wavelength (λ) (distance between the observed stripes), flow rate (Q), sample density (ρ_s), and mean radius of the jet (r) which is the average of maximum and minimum radii r_{max} and r_{min}:

$$\sigma = \frac{4}{6} \frac{\rho_s Q^2 \left[1 + \frac{37}{24} \left(\frac{r_{max} - r_{min}}{r_{max} + r_{min}} \right)^2 \right]}{r \lambda^2 \left[1 + \frac{5}{3} \pi^2 \frac{r^2}{\lambda^2} \right]} \tag{6.34}$$

Example 6.4. The capillary tube method was used to measure the surface tension of liquid helium-4 which has a density of 145 kg/m^3. A tube of radius 1.1 mm was immersed to a depth of 3.56 cm and the pressure was read as 51.3 Pa. If the contact angle is 0°, calculate the surface tension of liquid helium-4.

Solution:

Using Eq. (6.6):

$$\sigma = \frac{\Delta P r_t}{2 \cos \theta}$$

For 0° contact angle $\cos \theta = 1$

$$\Delta P = P - \rho g h$$

Then:

$$\sigma = \frac{(P - \rho g h) r_t}{2}$$

$$\sigma = \frac{\left[51.3 \, \text{Pa} - (145 \, \text{kg/m}^3)(9.81 \, \text{m/s}^2)(3.56 \times 10^{-2} \, \text{m}) \right] (1.1 \times 10^{-3} \, \text{m})}{2}$$

$$\sigma = 3.63 \times 10^{-4} \, \text{N/m}$$

Example 6.5. Pressure (gauge) of 2.91 MPa was necessary to blow water out of a glass filter with uniform pore width. What will be the mass of the drops leaving the filter at 20°C if the surface tension at that temperature is 0.07275 N/m? The density of water at 20°C is 997 kg/m^3 and the contact angle is 0°.

Solution:

Using Eq. (6.6):

$$\Delta P = \frac{2 \sigma \cos \theta}{r_t} \tag{6.6}$$

$$\cos \theta = 1$$

Then:

$$\Rightarrow r_t = \frac{2\left(0.07275\ \frac{N}{m}\right)(1)}{2.91 \times 10^6\ Pa} = 5 \times 10^{-8}\ m$$

The volume of the spherical drop can be calculated as:

$$
\begin{aligned}
V &= \frac{4}{3}\pi r^3 \\
&= \frac{4}{3}\pi \left(5 \times 10^{-8}\right)^3 \\
&= 5.23 \times 10^{-22}\ m^3
\end{aligned}
$$

Multiplying the determined volume by the density of the drop, the mass of the drop is found as:

$$
\begin{aligned}
m &= \rho V \\
&= \left(997\ \frac{kg}{m^3}\right)\left(5.24 \times 10^{-22}\ m^3\right) \\
&= 5.22 \times 10^{-19}\ kg
\end{aligned}
$$

PROBLEMS

6.1. A water droplet has a radius of 2×10^{-8} m. The surface tension of water at 30°C is 0.07 N/m. If the vapor pressure of a flat surface of water is 3160 Pa at 30°C, calculate the vapor pressure of water at that temperature.

6.2. Using a capillary rise method, the surface tension of olive oil was measured as 33×10^{-3} N/m. The height of rise in capillary glass tube that has a radius of 0.05 mm is 0.1 m. If the oil density is 800 kg/m^3, calculate the contact angle.

6.3. To find the degree of supersaturation required to nucleate water droplets spontaneously, water vapor is rapidly cooled to 20°C. It is found that the vapor pressure of water must be five times its equilibrium vapor pressure. If the surface tension of water is 73×10^{-3} N/m, calculate the radius of the water droplet formed at this degree of supersaturation.

6.4. Calculate the surface tension of milk if the height of rise of milk in a capillary of radius 0.001 cm is 30 cm and the contact angle is 70° ($\rho_{milk} = 1030$ kg/m^3).

6.5. An open glass tube that has a radius of 1 mm is inserted vertically into a dish of mercury at 20°C. Is there a rise or depression in the capillary? Calculate the amount of rise or depression. **Data:** Density of mercury: 13,600 kg/m^3, surface tension of mercury: 4.66×10^{-1} N/m, $\theta = 130°$

REFERENCES

Adamson, A.W. (1990) *Physical Chemistry of Surfaces*, 5th ed. New York: Wiley-Interscience.

Arnebrant, T., & Naylander, T. (1985). Surface tension measurements by an automated drop volume apparatus. *Journal of Dispersion Science and Technology*, 6, 209–212.

Aveyard, R. & Haydon, D.A. (1973). *An Introduction to the Principles of Surface Chemistry*. Cambridge, UK: Cambridge University Press.

Bertsch, A.J. (1983). Surface tension of whole and skim milk between 18 and 135°C. *Journal of Dairy Research, 50*, 259–267.

Bos, M.A., & van Vliet, T. (2001). Interfacial rheological properties of adsorbed protein layers and surfactants: A review. *Advances in Colloid and Interface Science, 91*, 437–471.

Choi, W.Y., Park, H.J., Ahn, D.J., Lee, J., & Lee, C.Y. (2002). Wettability of chitosan solution on 'Fuji' apple skin. *Journal of Food Science, 67*, 2668–2672.

Coia, K.A., & Stauffer, K.R. (1987). Shelf life study of water/oil emulsions using various hydrocolloids. *Journal of Food Science, 52*, 166–175.

Dann, J.R. (1970). Forces involved in the adhesive process. I. Critical surface tensions of polymeric solids as determined with polar liquids. *Journal of Colloid and Interface Science, 32*, 302–331.

Dickinson, E. (1992) *An Introduction to Food Hydrocolloids*. Oxford, UK: Oxford University Press.

Dickinson, E. (2003). Hydrocolloids at interfaces and the influence on the properties of dispersed systems. *Food Hydrocolloids, 17*, 25–39.

Duthie, A.H., & Jenson, R.G. (1959). Influence of added 1-monoglycerides on the surface tension of milk. *Journal of Dairy Science, 42*, 863.

Flingoh, C.H.O.H., & Chong Chiew, L. (1992). Surface tension of palm oil, palm olein and palm stearin. *Elaeis: Journal of the Palm Oil Research Institute of Malaysia, 4*, 27–31.

Fowkes, F.M. (1964). Attractive forces at interfaces. *Industrial and Engineering Chemistry, 56*, 40–52.

Fowkes, F.M. (1972). Donor-acceptor interactions at interfaces. *Journal of Adhesion, 4*, 155–159.

Fowkes, F.M. (1983). Acid-base interactions in polymer adhesion. In K.L. Mittal (Ed.), *Physicochemical Aspects of Polymer Surfaces* (pp. 583–603), New York: Plenum Press.

Garti, N., & Leser, M.E. (2001). Emulsification properties of hydrocolloids. *Polymers for Advanced Technologies, 12*, 123–135.

Girifalco, L.A., & Good, R.J. (1957). A theory for the estimation of surface and interfacial energies. I. Derivation and application to interfacial tension. *Journal of Physical Chemistry 61*, 904–909.

Ikegami, S., Williams P.A., & Phillips, G.O. (1992). Interfacial properties of xanthan gum. In G.O. Phillips, D.J. Wedlock & P.A. Williams (Eds.), *Gums and Stabilizers for the Food Industry* (Vol. 2, pp. 371–377). Oxford: IRL Press.

James, A.M. (1967). *Practical Physical Chemistry*, 2nd ed. London: J & A Churchill.

Jenness, R., Shipe, W.F, Jr., & Sherbon, J.W. (1974). Physical Properties of milk. In B.H. Webb, A.H. Johnson & J.A. Alford (Eds), *Fundamentals of Dairy Chemistry*. Westport, CT: AVI.

Kaye, G.W.C., & Laby, T.H. (1973). *Table of Physical and Chemical Constants*. London: Longmans.

Kokelaar, J.J., Van Vliet, T., & Prins, A. (1994). Surface and bulk properties in relation to bubble stability in bread dough. In E. Dickinson & D. Lorient (Eds.), *Food Macromolecules and Colloids* (pp. 277–284). Cambridge: Royal Society of Chemistry.

Krog, N., & Lauridsen, J.B. (1977). Food emulsifiers and their associations with water. In S. Frieberg (Ed.), *Food Emulsions* (pp. 67–139). New York: Marcel Dekker.

Kubiak A., & Dejmek, P. (1993). Application of image analysis to measurement of dynamic surface tension using oscillating jet method. *Journal of Dispersion Science &. Technology, 14*, 661–673.

Lewis, M.J. (1996) *Physical Properties of Foods and Food Processing Systems*. Cambridge, UK: Woodhead.

Li, D., & Neumann, A.W. (1992). Equation of state for interfacial tensions of solid-liquid systems. *Advance in Colloid and Interface Science, 39*, 299–345.

McGuire J. (2005). Surface properties. In M.A. Rao, S.S.H. Rizvi & A.K. Datta (Eds.), *Engineering Properties of Foods*, 3rd ed. (pp. 679–702). Boca Raton: CRC Press Taylor & Francis.

Michalski, M.C., Desobry, S., & Hardy, J. (1997). Food materials adhesion: A review. *Critical Reviews in Food Science and Nutrition, 37*, 591–619.

Ould-Eleya, M., & Hardy, J. (1993). Evaluation of the food adhesion onto packaging materials. In *Food Preservation 2000: Integrating, Processing and Consumer Research*, October 19 to 21, Massachusetts: US Army Natick Research.

Peron, N., Cagna, A., Valade, M., Marchal, R., Maujean, A., Robillard, B., Aguie-Beghin, V., & Douillard, R. (2000). Characterization by drop tensiometry and by elipsometry of the adsorption layer formed at the air/champagne wine interface. *Advances in Colloid and Interface Science, 88*, 19–36.

Powrie, W.D., & Tung, M.A. (1976). Food Dispersions. In O.R. Fennema (Ed.), *Principles of Food Science, Part I, Food Chemistry*, (pp. 539–575). New York: Marcel Dekker.

Riedl, Z., Szklenarik, G., Zelko, R., Marton, S., & Racz, I. (2000). The effect of temperature and polymer concentration on dynamic surface tension and wetting ability of hydroxymethylcellulose solutions. *Drug Development and Industrial Pharmacy, 26,* 1321–1323.

Sasaki,K., Shimaya, Y., Hatae, K., & Shimada, A. (1991). Surface tension of baked food batter measured by the maximum bubble pressure method. *Agricultural & Biological Chemistry, 55,* 1273–1279.

Shaw D. (1970). *Introduction to Colloidal and Surface Chemistry,* 2nd ed. London: Butterworths.

Van Oss, C.J., Good R.J, & Busscher, H.J. (1990). Estimation of the polar surface tension parameters of glycerol and formamide for use in contact angle measurements on polar solids. *Journal of Dispersion Science and Technology, 11,* 75–81.

Walstra, P. (1983). Formation of emulsions. In P. Becher (Ed.), *Encyclopedia of Emulsion Technology* (Vol. 1, pp. 57–127). New York: Marcel Dekker.

Walstra, P., &. Smulders, P.E.A. (1998). Emulsion formation. In B.P. Binks (Eds.), *Modern Aspects of Emulsion Science* (pp. 56–99). Cambridge, UK: Royal Society of Chemistry.

Watson, P.D. (1956). The effect of variations in fat and temperature upon the surface tension of various milks. *Journal of Dairy Science, 39,* 916.

Watson, P.D. (1958). Effect of variations in fat and temperature upon the surface tension of various milks. *Journal of Dairy Science, 41,* 1693–1697.

Weast R.C. (1982). *Handbook of Physics and Chemistry.* 63rd ed. Cleveland, OH: CRC Press.

Witnah, C.H. (1959). The surface tension of milk: A review. *Journal of Dairy Science, 42,* 1437–1449.

Xu, W., Nikolov, A., Wasan, D.T., Gonsalves, A., & Borwankar, R.P. (2003). Foam film rheology and thickness stability of foam-based food products. *Colloids and Surfaces A-Physicochemical and Engineering Aspects, 214,* 13–21.

Index